Molecular Simulation on Cement-Based Materials

Dongshuai Hou

Molecular Simulation on Cement-Based Materials

From Theory to Application

Dongshuai Hou
Civil Engineering
Qingdao Technological University
Qingdao, China

ISBN 978-981-13-8713-5   ISBN 978-981-13-8711-1   (eBook)
https://doi.org/10.1007/978-981-13-8711-1

Jointly published with Science Press
The print edition is not for sale in China. Customers from China please order the print book from:
Science Press

This Springer imprint is published by the registered company Springer Nature Singapore Pte Ltd.
The registered company address is: 152 Beach Road, #21-01/04 Gateway East, Singapore 189721,
Singapore

Foreword

Portland cement concrete is the most widely used building material in the world. It plays an important role in infrastructure and private buildings' construction. Concrete is basically made of filler and Portland cement. It is the hydration products of Portland cement that glue the individual particles of the filler together to form a solid material. Obviously, the mechanical property enhancement of concrete needs a deep understanding of the nature of hydration product. Especially, a quantitative understanding of calcium silicate hydrate (C–S–H), the most important hydration product, on the atomic to 100 nm scale is one of the most important, long-standing needs in cement science as it can be used to explain how to control the mechanical, transport, and chemical properties of hydrated cement paste. Fortunately, with the advancement in technology, a wide range of experimental and computational tools are available in recent decades for discovering the nature of hydrate of cementitious materials.

The numerical simulation methods at atomic and molecular scale include quantum chemical and molecular potential-based methods. Quantum (first principles) approaches involve solution of the Schrödinger equation describing the interaction of electrons and atomic nuclei. The electrons are described by their wave functions, and the challenge is to adequately describe these functions in a computationally accessible way. The Schrödinger equation cannot be solved exactly except in the most limited cases. Hence, some approximate methods are needed. The density function theory provides an approximate solution of the exact solution to an approximate Schrödinger equation.

Potential-based methods, such as Molecular Dynamics (MD) approach, are based on empirical or semi-empirical potentials between or among atomic or molecular entities. They involve treating the atoms or molecules as classical (non-quantum) entities and computing their positions, motion, and energies as they interact with each other under the influence of potential functions. These functions can describe short range atomic repulsion, van der Waals forces, and attractive and repulsive Coulombic interactions. MD methods follow the time evolution of the structure and energy of the computed system, thus allowing calculation of dynamical properties such as vibrational spectra. This book provides a more detailed discussion of applications of potential-based, molecular dynamics simulations to cement systems. It summarizes the author's more than ten years of experiences using MD simulation to reveal the intrinsic mechanism for cement-based materials and to guide the scientific design of cementitious materials. The book is aimed at providing handy application methodology on MD principles for the readers in civil engineering field and thus introduced the simulation techniques such as potential forms and energy minimization as well as the related common experiment methods to cement-based materials with logical thinking in civil engineering.

The unique feature of this book is directly related to the points of cement and concrete chemistry from molecular dynamics modeling. It provides many useful examples of cement hydrate simulation such as ion and water migration in C–S–H gel, interfacial behavior of solution species in the nanometer-scaled channel, and mechanical properties of C–S–H gel. These examples are very practical and useful for civil engineers.

The book is divided into eight chapters. Chapter 1 gives a brief introduction of background and objective of the book. Chapter 2 provides the basic knowledge of modeling of cement hydrate at nano scales. Chapter 3 discusses the molecular mechanics and dynamics. Chapter 4 focuses on modeling of C–S–H by MD, including the mechanical properties of calcium silicate hydrate at nanometer scale. Chapter 5 covers the simulation on water and ions' migration in the nanometer channel. Chapter 6 provides updated knowledge on models for the cross-linked calcium-aluminate-silicate-hydrate gel. Chapter 7 introduces the molecular dynamics study on cement-graphene nanocomposite. Chapter 8 discusses the issues regarding the future and development trend of computational chemistry applied in concrete science.

The book is designed and written primarily to meet the needs for teachers and graduate students' researches which are involved in the characterization of materials' structure in nanometer scale of concrete hydrate in civil engineering. However, it can serve as a reference or a guide for professional engineers in their practice.

The author of this book is a new star in civil engineering materials, especially in MD simulation for cement-based materials. Wish this book makes a new milestone in his career.

Macao, China   Zongjin Li
August 2019   University of Macao

Preface

Concrete, the largest material by volume used by mankind, is ubiquitously utilized in the construction industry. Currently, concrete production and manufacture contribute to about 6–8% of yearly man-made global greenhouse gas emissions. One important strategy to reduce the carbon footprint induced by concrete is to produce more durable and sustainable cement-based material. Understanding the intrinsic behaviors of concrete at the atomistic level is essential for high-performance concrete material design. Even though the atomic properties of the cement-based material have been explored by experiments for many years, the molecular properties of calcium silicate hydrate (C–S–H), the essential hydration product of cement hydrate, has not been comprehensively understood. The motivation to write this book is to introduce the computational chemistry method to help decode the molecular structure of calcium silicate hydrate and give a valuable complementary interpretation of the experimental studies. This book also provides more comprehensive knowledge on the molecular dynamics (MD) simulation, including the systematic introduction of the theory of force field, the MD algorithm and MD trajectory evaluation methods.

The unique feature of his book includes many aspects of cement and concrete chemistry that benefit from molecular dynamics modeling. The computational chemistry method helps reveal the intrinsic mechanism of the material, and give a scientific guide for the material design. The migration of ions and water in the C–S–H gel, influencing cohesive strength of the hydrate, determines the durability of concrete material. The molecular modeling reveals the structure, dynamics, and interfacial behavior of solution species ultra-confined in the nanometer channel. Based on the molecular dynamics model, the molecular structure evolution as the function of chemical composition is investigated, and the structure–property correlation of cement hydrate is established at the molecular level for optimization in material design. This book also includes molecular dynamics investigation on the cement hydrate modified by nanomaterials such as the graphene oxide sheet. This contributes to advancement to produce more ductile cement-based material by nanotechnology. The successful application of MD in cement material investigation confirms the great power and promising future of computational method in the field of concrete science.

In the process of writing this book, the authors received enthusiastic help and invaluable assistance from many people, which is deeply appreciated. The authors would like to express his special thanks to Dr. Pan Wang, Dr. Jiao Yu, Mr. Yu Zhang, Mr. Tao Li, Mr. Jun Yang, Mr. Tiejun Yang, Miss Xiaoqian Xu, Mr. Gang Qiao, and Mr. Wei Zhang, for their help in editing the book draft and professional page proofreading.

The support from China Basic Research Grant, under Grant 51508292, 51678317, 51420105015, Fok Ying Tung Education Foundation under Grant 161069, the China Ministry of Science and Technology under Grant 2015CB655100, Natural science foundation of Shandong Province under Grant ZR2017JL024, are greatly acknowledged.

Finally, I would like to thank for my wife, Mengmeng Li, my daughters Ling-yun Hou and Ling-xuan Hou for their love, understanding, and support.

Qingdao, China   Dongshuai Hou

Contents

Chapter 1
Background and Objectives

1.1 Introduction

Concrete, the most widely used building material, has been applied to make pavements, architectural structures, foundations, motorways, roads, overpasses, parking structures, brick, block walls and footings for gates, fences, and poles. And its usage quantity is incomparable by other materials [1]. Hence, the usage of concrete symbolizes the development level of modern urbanization and closely relates to the life of human beings. But this does not mean that concrete is a perfect material. As a matter of fact, there are still many problems in concrete material that needed to be solved [2–5]. First, the concrete contains its intrinsic defects: (1) the tensile and flexural strength of concrete is much lower compared to its compressive strength. Generally, the tensile and flexural strength for concrete are of only 10% and 15%, respectively, of its compressive strength [6]. (2) The hydration of concrete would inevitably cause chemical shrinkage [7]. Besides, there are also autogenous shrinkage, air shrinkage, and creep of concrete, which would happen at certain circumstances. Second, the manufacture of Portland cement, the core constituent of concrete, is an energy-consuming industry. About 0.85 ton of CO_2 is released for one ton of Portland cement production [8]. Hence, the increasing demand of concrete poses great threat to the natural environment. Third, concrete material is susceptible to the attack of aggressive ions from environment [9, 10]. In marine or brine service environment, the attack of aggressive ions would lead to the expansion and cracking or even loss of cohesive strength for the concrete [11]. The premature failure of concrete structure would, in turn, demand for more cement production and exert larger environmental effect. Therefore, many measures have been performed in preparing concrete to endow it with new functions or reduce its ecological footprint [12, 13]. For example, the combination of concrete with rebar or steel fiber to improve its tensile and flexural strength, the incorporation of expansion agent to compensate its shrinkage, the addition of mineral admixture or polymer to improve its durability, and even the full replacement of Portland cement by supplementary cementitious materials (SCMs) to achieve sustainability in construction industry. Owing to these modifications, the

© Science Press and Springer Nature Singapore Pte Ltd. 2020
D. Hou, *Molecular Simulation on Cement-Based Materials*,
https://doi.org/10.1007/978-981-13-8711-1_1

traditional "Portland cement concrete" also becomes general "cement-based materials". However, the abovementioned measures can only modify the performance of cement-based materials to some extent or to some aspects. With the increasing understanding of the cement-based materials, the investigations and modifications on cement-based materials down to the nanoscale are still going on.

1.2 Research Motivation of Atomistic-Scale Simulation

Nanotechnology refers to the comprehending and manipulation of matter on the order of nanometers to produce materials with new properties and functions. After the famous lecture "There's Plenty of Room at the Bottom" by Richard P. Feynman in 1959, many revolutionary developments have been achieved in physics, chemistry, and biology through nanotechnology, which proves the prediction of Feynman [14]. Similarly, the investigation and modification of cement-based materials at the nanolevel is a prerequisite for the fundamental improvement of their properties and performance. The cement paste is a porous and hierarchical material with different phases. When cement powder is mixed with water, the hydration reaction gives birth to main chemical products: Calcium silicate hydrate (C–S–H), calcium hydroxide (CH), ettringite, and monosulphoaluminates. C–S–H takes up approximately 50–70% of the fully hydrated products and contributes predominantly to the mechanical strength of the cement-based material [15, 16]. Therefore, the understanding and control of C–S–H gel is the key to the revolutionary improvement in cement-based materials. However, due to the hierarchical nature of the concrete, the concrete structure has various levels at different length scales. As shown in Fig. 1.1, at macro-level ($>10^{-3}$ m), the concrete can be taken as a material composed mainly by two phases: the aggregate and the cement matrix. At micro-meter level (10^{-3}–10^{-6} m), various hydration products and the different morphologies of the C–S–H gel form the heterogeneous structure of the cement paste. At meso-scale, less than 1 μm, the C–S–H gel can be observed as ellipsoid-shaped particles packing with different densities. At nanometer scale ($<10^{-9}$ m), the distribution and arrangement of Ca, Si, OH form the C–S–H basic structure. Due to the compositional, structural, and physical complexity of the C–S–H gel, the basic building elements of cement hydrate, has not been understood comprehensively yet [17].

The first research aim of the book is to solve the intrinsic molecular structure of C–S–H gel at nanolevel. An accurate description of the C–S–H structure should not only be consistent with experimental observations, but also comply with the basic principles of crystal chemistry and physics. Since the backbone of C–S–H gel can be taken as the DNA of concrete structure, it is necessary and urgent to decode the structure of C–S–H gel. Based on the molecular structure of the C–S–H gel, the interplay between the composition, structure, mechanics, and transport properties is further investigated. The mechanical property of the C–S–H gel is essential because the mechanical behavior at nano- and microscale is the cohesive resource of the material at macro-level. Water and ions transport in the C–S–H gel at nanoscale determines

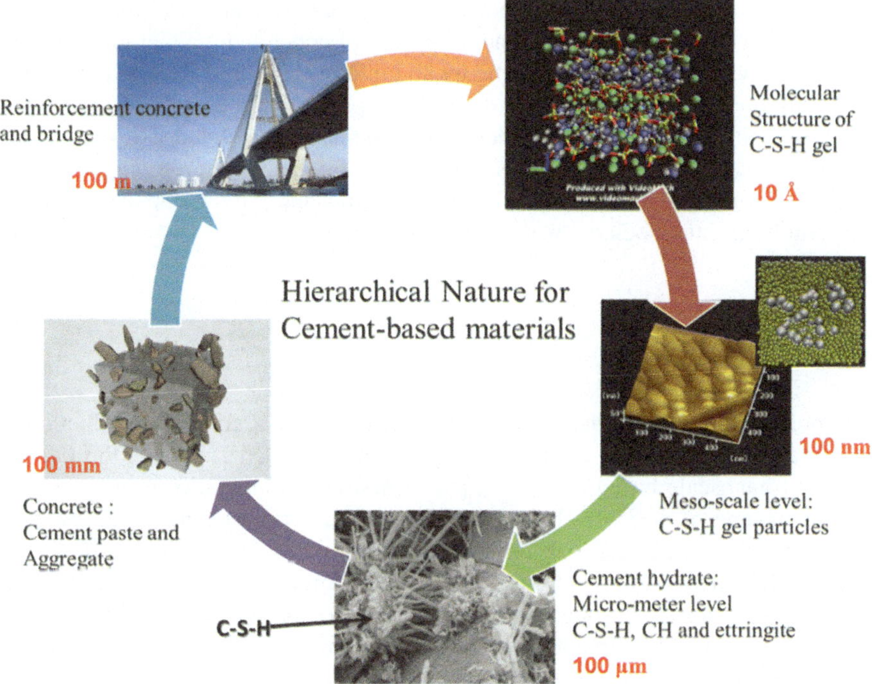

Fig. 1.1 Hierarchical structure of concrete

the long-term durability of the cement-based materials. Hence, the transport of water and ions in the as-constructed C–S–H models are also studied. Besides, this book investigates two types of modified C–S–H gel at the nanometer scale utilizing molecular dynamics simulation. The first one is the aluminum-induced cross-links on the C–A–S–H gel. Due to the large quantity of aluminum-rich SCMs (e.g., fly ash and granulated blast furnace slag) used in preparing concrete, the cross-linked C–A–S–H is ubiquitous and worth of researching [18]. The second one is the combination of C–S–H gel with graphene/graphene oxide [19, 20]. Combination of C–S–H gel, an inorganic material, with polymers at the atomic level can give radically new insights into the cement-based materials.

All of these behaviors related with cement chemistry at the nanoscale level and the mechanisms behind the abovementioned points have been poorly understood even if some knowledge of the microscopic-level chemistry is well documented. To be able to predict the intrinsic origin of properties, durability and degradation of concrete material, a concerted bottom-up approach, with an effort at both the atomistic scale to give accurate inputs into microstructural models that are already well developed. With this better understanding of atomistic-scale mechanisms, experimental work, and material design at the larger concrete testing scale will be better directed and more efficient, eliminating much of the "test and try" methodology.

Continually increasing computing power has allowed computational materials science to predict the microstructure and properties of concrete by integrating physics and chemistry. Atomic-scale modeling by ab initio, molecular dynamics (MD), and energy minimization techniques has offered new potentials in cement science and is shedding light onto the processes occurring at the nanoscale that are of importance for understanding material properties at the macro-level and for developing strategies to improve concrete performance. This will help to achieve a satisfactory reduction in carbon footprint associated with cementitious systems.

1.3 Outline of Book

The book contains eight chapters. Chapters 2 and 3 give a general introduction to the cement and molecular modeling, respectively. Chapter 2 reviews the C–S–H models at nanoscale based on both experimental and computational method. The intrinsic features of the C–S–H gel and the challenge in model construction are also discussed. Chapter 3 deals with introducing the basic principles of the simulation method. The reactive force field molecular simulation is the major approach in the model construction and properties calculation. Based on the experimental characterization and simulation method in previous chapters, new C–S–H models with the different compositions (Ca/Si and Water/Ca ratios) are proposed in Chap. 4. These models, improving structural defects of previous cement model, are not only consistent with the experiment observation but also comply with the crystal chemistry. Furthermore, the influence mechanisms of composition on the structural and mechanical properties of the C–S–H gel are also studied. In Chap. 5, the water and ions adsorption and capillary transport in the nanopores of calcium silicate hydrate. The transport of water and ions influences significantly the durability of the C–S–H gel. Hence, the obtained results are helpful in interpreting the influence of composition on the durability of cement-based materials. In Chap. 6, the cross-linked C–A–S–H gel induced by aluminum incorporation is investigated in respect to its structure, dynamics, reactivity, and mechanical properties. In Chap. 7, the intrinsic interaction between the graphene/GO and cement hydrate in high-performance cement-based composites has been investigated at the atomic, micro, and macrolevels. Chapter 8 provides promising perspectives on the application of the molecular model in cement and concrete research in future.

References

1. Li, Z. (2011). *Advanced Concrete Technology*.
2. Barcelo, L., Kline, J., Walenta, G., & Gartner, E. (2014). Cement and carbon emissions. *Materials and Structures, 47*(6), 1055–1065.

3. Gartner, E., & Hirao, H. (2015). A review of alternative approaches to the reduction of CO_2 emissions associated with the manufacture of the binder phase in concrete. *Cement and Concrete Research, 78,* 126–142.
4. Lollini, F., Redaelli, E., & Bertolini, L. (2016). A study on the applicability of the efficiency factor of supplementary cementitious materials to durability properties. *Construction and Building Materials, 120,* 284–292.
5. Mehta, P. K. (1997). Durability—Critical issues for the future. *Concrete International, 19*(7), 27–33.
6. Monteiro, P. (2006). *Concrete: Microstructure, properties, and materials.* McGraw-Hill Publishing.
7. Persson, B. (1998). Experimental studies on shrinkage of high-performance concrete. *Cement and Concrete Research, 28*(7), 1023–1036.
8. Feiz, R., Ammenberg, J., Baas, L., Eklund, M., Helgstrand, A., & Marshall, R. (2015). Improving the CO_2, performance of cement. Part I: Utilizing life-cycle assessment and key performance indicators to assess development within the cement industry. *Journal of Cleaner Production, 98,* 272–281.
9. Neville, A. (2004). The confused world of sulfate attack on concrete. *Cement and Concrete Research, 34*(8), 1275–1296.
10. Hobbs, D. W., & Taylor, M. G. (2000). Nature of the thaumasite sulfate attack mechanism in field concrete. *Cement and Concrete Research, 30*(4), 529–533.
11. Shi, Z., Lothenbach, B., Geiker, M. R., Kaufmann, J., Leemann, A., Ferreiro, S., et al. (2016). Experimental studies and thermodynamic modeling of the carbonation of Portland cement, metakaolin and limestone mortars. *Cement and Concrete Research, 88,* 60–72.
12. Mao, J., Jin, W., Zhang, H., Chen, X. U., & Xia, J. (2015). Technology for enhancing durability of structures of sea-sand concrete and its application. *Journal of Chinese Society for Corrosion & Protection, 35*(6).
13. Bernal, S. A., Rodríguez, E. D., Kirchheim, A. P., & Provis, J. L. (2016). Management and valorisation of wastes through use in producing alkali-activated cement materials. *Journal of Chemical Technology and Biotechnology, 91*(9), 2365–2388.
14. Sanchez, F., & Sobolev, K. (2010). Nanotechnology in concrete—A review. *Construction and Building Materials, 24*(11), 2060–2071.
15. Richardson, I. G. (2008). The calcium silicate hydrates. *Cement & Concrete Research, 38*(2), 137–158.
16. Mirzahosseini, M., & Riding, K. A. (2015). Influence of different particle sizes on reactivity of finely ground glass as supplementary cementitious material (SCM). *Cement & Concrete Composites, 56,* 95–105.
17. Allen, A. J., Thomas, J. J., & Jennings, H. M. (2007). Composition and density of nanoscale calcium–silicate–hydrate in cement. *Nature Materials, 6,* 311.
18. Hong, S. Y., & Glasser, F. P. (2002). Alkali sorption by C–S–H and C–A–S–H gels: Part II. Role of alumina. *Cement & Concrete Research, 32*(7), 1101–1111.
19. Pan, Z., He, L., Qiu, L., Korayem, A. H., Li, G., Zhu, J. W., et al. (2015). Mechanical properties and microstructure of a graphene oxide–cement composite. *Cement & Concrete Composites, 58,* 140–147.
20. Chong, Y. R., Chang, E. K., Lee, C. S., Kim, K. I., & Lee, S. K. (1999). Preparation and characterization of absorbent polymer-cement composites. *Cement and Concrete Research, 29*(2), 231–236.

Chapter 2
Introduction to Modeling of Cement Hydrate at Nanoscale

The general background on the cement hydrate is introduced in this chapter. The characteristics of the C–S–H gel obtained by various experimental techniques are summarized firstly. This chapter also emphasizes on a series of theoretical models of C–S–H gel at nanoscale. Both the experimental and theoretical information provide the foundation for C–S–H model construction.

2.1 Formation of the C–S–H Gel

Hydration of the cement is the reaction between the dry cement powder and water solution. Portland cement has four major compounds: tri-calcium silicate (C_3S), di-calcium silicate (C_2S), tri-calcium aluminate (C_3A), and tetra-calcium alumino-ferrite (C_4AF). The hydration product for aluminate phase is the ettringite, which is the needle-shaped crystal. The C_3S and C_2S, comprising over 80% by weight of most cement, are responsible for the main composition of the hydration products. While the C_3S is the most important phase in cement for strength development during the first month, the C_2S reacts much more slowly, and contributes to the long-term strength of the cement. Both the silicate phases react with water as shown below to form calcium hydroxide and a rigid calcium-silicate hydrate gel, C–S–H gel.

$$2(3CaO \cdot SiO_2) + 6H_2O = 3CaO \cdot 2SiO_2 \cdot 3H_2O + 3Ca(OH)_2 \tag{2.1}$$

$$2(2CaO \cdot SiO_2) + 4H_2O = 3CaO \cdot 2SiO_2 \cdot 3H_2O + Ca(OH)_2 \tag{2.2}$$

C–S–H gel is the principle hydration product, occupying more than 60% of the structural component in the cement paste. It is believed that C–S–H is responsible for the strength, shrinkage, and durability of the Portland cement concrete. However, due to diversity of the Ca/Si ratio and structural water content, the real composition of the C–S–H gel is not as simple as proposed in the chemical formulas. It can be

© Science Press and Springer Nature Singapore Pte Ltd. 2020
D. Hou, *Molecular Simulation on Cement-Based Materials*,
https://doi.org/10.1007/978-981-13-8711-1_2

clearly seen the morphology evolution of high w/s ratio hydration from Fig. 2.1. Some hydration products are formed on the surface of original C_3S particles. There are mainly three different kinds of morphology: (i) the rectangular-like one with clear edges and corners, which are the crystal of calcium hydroxide, (ii) bundles of fibrillar-like one, and (iii) ball-like one. The diversity of the morphologies implies the complexity of C–S–H structure.

Even though the composition and the structure of C–S–H gel have not been comprehensively understood, the cement hydration kinetic is well known [2]. In the first few minutes of hydration (Fig. 2.2a), the aluminum-ferrite phases react with gypsum to form an amorphous gel at the surface of the cement grains and short rods of ettringite grow. Cement hydration slows down and the induction period begins after the initial reaction. Subsequently, the acceleratory period follows the induction period. During the period from 3 to 24 h, about 30% of cement particles react to

(a)   **(b)**

(c)   **(d)**

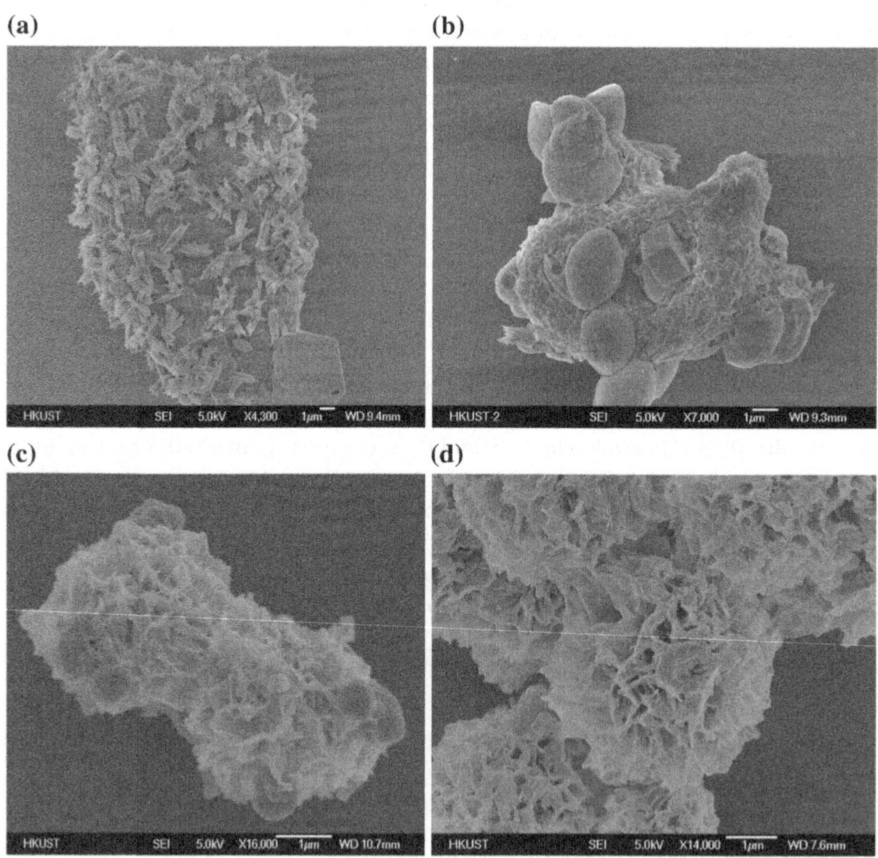

Fig. 2.1 SEM of C_3S hydration product under different hydration time **a** 1 day, **b** 7 days, **c** 14 days, **d** 30 days [1]

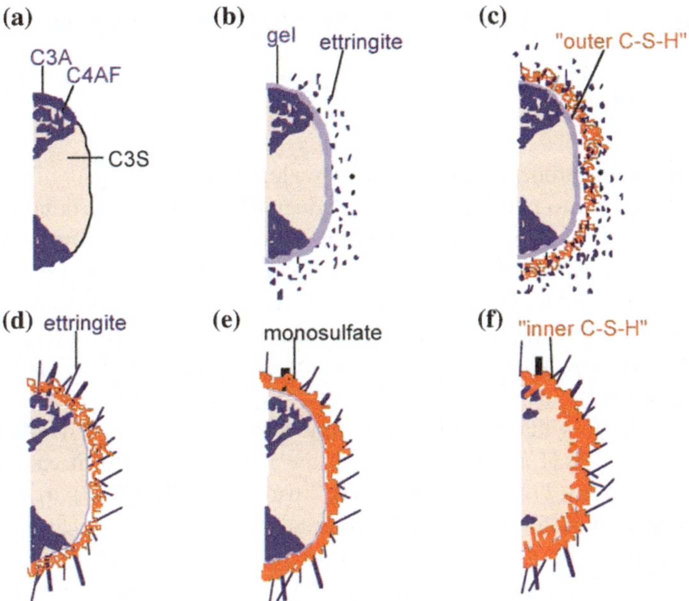

Fig. 2.2 Schematic representation of anhydrous cement (**a**) and the effect of hydration after **b** 10 min, **c** 10 h, **d** 18 h, **e** 1–3 days, and **f** 2 weeks [3]

form calcium hydroxide and C–S–H gel. The development of C–S–H in this period occurs in 2 phases. After 10 h hydration (Fig. 2.2c), "outer C–S–H," grows out from the ettringite rods rather than directly out from the surface of the C_3S particles. In the latter part of the acceleratory period, after 18 h of hydration, C_3A continues to react with gypsum, forming longer ettringite rods (Fig. 2.2d). This network of ettringite and C–S–H appears to form a "hydrating shell" about 1 μm from the surface of anhydrous C_3S. The "inner C–S–H" gel forms inside this shell. After 1–3 days of hydration, reactions slow down and the deceleratory period begins (Fig. 2.2e). C_3A reacts with ettringite to form some mono-sulfate. "Inner C–S–H" continues to grow near the C_3S surface, narrowing the 1 μm gap between the "hydrating shell" and anhydrous C_3S. The rate of hydration is likely to depend on the diffusion rate of water or ions to the anhydrous surface. After 2 weeks hydration (Fig. 2.2f), the gap between the "hydrating shell" and the grain is completely filled with C–S–H. The original, "outer C–S–H" becomes more fibrous.

2.2 Experimental Characterization of the C–S–H Gel

2.2.1 Morphology

From the hydration process, at micrometer scale, the morphologies of C–S–H gel can be categorized into two types: inner products (Ip) and outer products (Op). As shown in Fig. 2.3b, the Ip grows near the cement grains and has relative high density. It is composed of aggregated globule particles of 4–6 nm and the pore size inside the Ip is less than 10 nm. On the contrary, in Fig. 2.3c, for the Op, less space restriction and sufficient water environment can make the structure grow to fibril-like (length to width ratio is high) morphologies. This structure style can leave large porosity between the fibrils.

At mesoscale, less than 1 μm, the atomic force microscopy (AFM) can help characterize the C–S–H at this scale. From the AFM photos [5], it can be clearly observed in Fig. 2.4 the boundaries between ellipsoid shaped C–S–H particles and the packing orientation of those particles. Furthermore, recent series of nanoindentation test prove the granular mechanical properties of C–S–H particles [6] and results from the small angle neutron scattering (SANS) tests also indicate the existence of the nanoparticles aggregation [7] as shown in Fig. 2.4. Nevertheless, the shapes, sizes and packing distribution for the C–S–H particles are not determined quantitatively. The formation mechanism of the C–S–H colloid is up to solve.

At nanolevel ($<10^{-9}$), multi techniques such AFM, nuclear magnetic resonance (NMR), SANS, small angle X-ray pectroscopy (SAXS), have attempted to probe the molecular structure of CSH. Nevertheless, these methods cannot solve the structure straightforward and no consistent model can be proposed only by a single technique. These methods characterize the properties of the C–S–H in different aspects.

2.2.2 Ca/Si Ratios

An important stoichiometric parameter that defines a C–S–H phase is the molar ratio of CaO to SiO_2 in its structure (C/S ratio). It is possible to calculate the C/S ratio of the C–S–H produced in the hydration of calcium silicates by determining the content of calcium hydroxide and unreacted materials using analytical methods such as thermo-gravimetric analysis, quantitative X-ray diffraction and higher resolution TEM techniques.

Figure 2.5 shows a Ca/Si ratio frequency histogram for X-ray microanalyses in the TEM of C–S–H present in hardened Portland cement pastes aged 1 day to 3 1/2 years. This is a large set of data which illustrates the extent of compositional variation possible for C–S–H present in neat Portland cement pastes (hydrated at 20 °C; W/C = 0.4): from a Ca/Si ratio of 1.2 to 2.3 with a mean of 1.75. It should be noted that the distributions vary with age: C–S–H present in young Portland cement

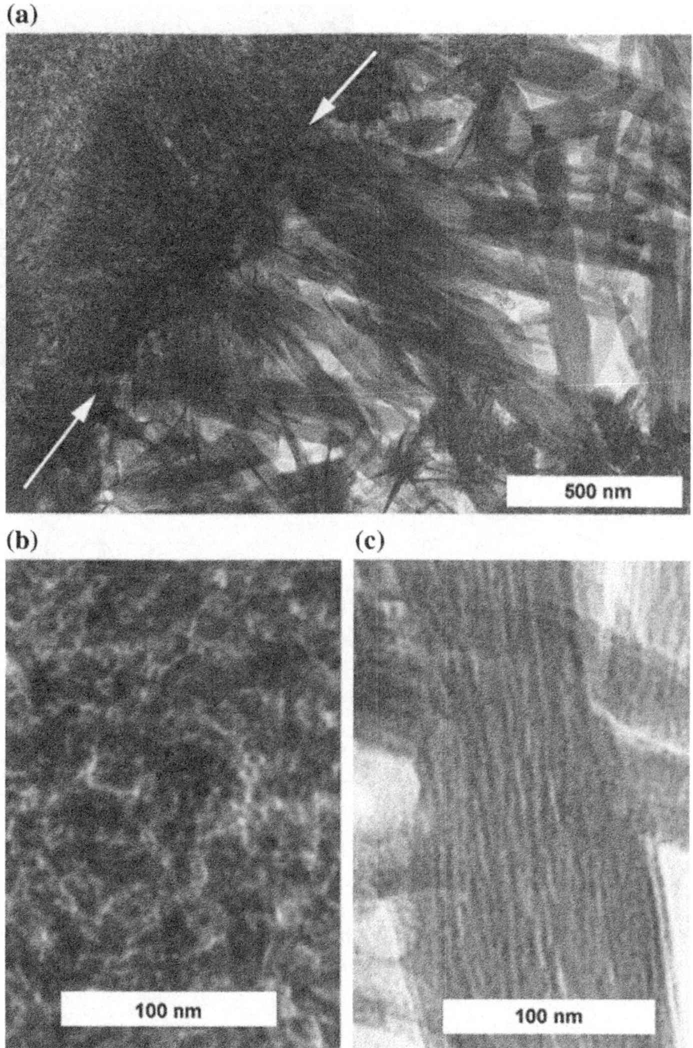

Fig. 2.3 **a** A TEM micrograph showing Ip and Op C–S–H present in a hardened C3S paste. White arrows indicate the Ip–Op boundary. **b** An enlargement of a region of Ip C–S–H. **c** An enlargement of a fibril of Op C–S–H [4]

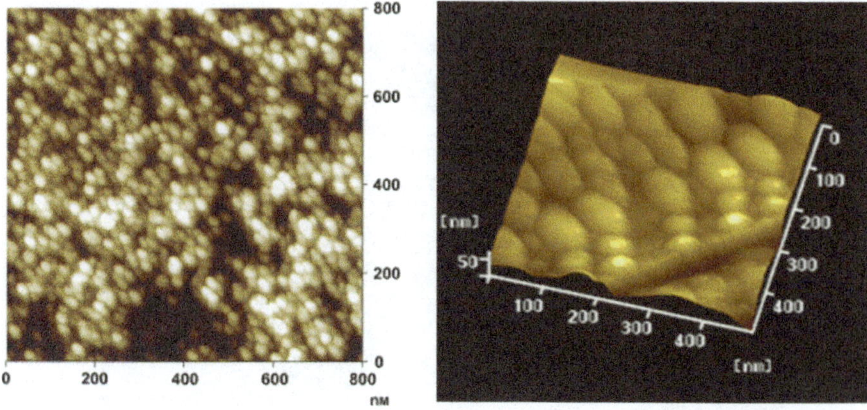

Fig. 2.4 AFM photos for the CSH particles [5]

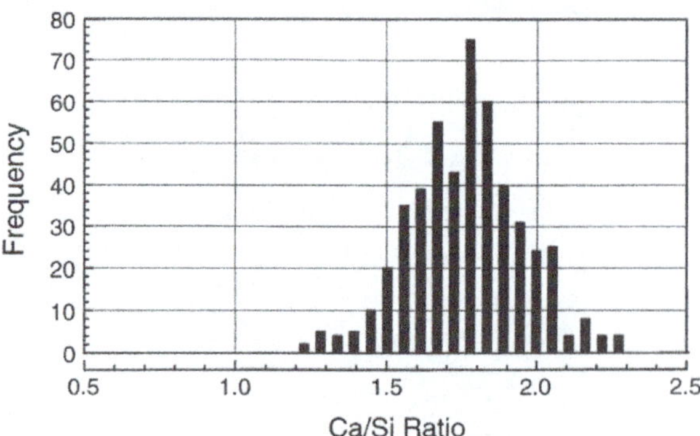

Fig. 2.5 Ca/Si ratio frequency histogram for C–S–H in Portland cement pastes aged 1 day to 3 1/2 years (493 TEM microanalyses of C–S–H free of admixture with other phases) [8]

pastes displays a bimodal Ca/Si ratio distribution which becomes unimodal with age. Ip C–S–H generally has a higher mean Ca/Si ratio than Op [9].

2.2.3 Water States

C–S–H gel has a multi-scale porous structure that contains capillary pores and gel pores. The water and ion diffusion in the C–S–H gel determines the strength, creep, shrinkage, chemical, and physical reactivity of cementitious materials. On the one hand, environmental changes, such as the temperature, humidity, and loading can result in water content variation in the C–S–H gel. The interaction between water and C–S–H gel can significantly affect the cohesion force [10]. On the other hand, in an aggressive environment, such as in seawater, the Cl^- ions can penetrate into the cement paste and lead to the corrosion of the reinforcement [11, 12], and brings about detrimental effects on the durability of the structure. In order to interpret the diffusion mechanism at different length scales, it is necessary to investigate the origin of the properties at the molecular level.

The properties of the water confined in gel pore or in the vicinity of C–S–H surface have been studied by various experimental techniques. By using the ^{1}H nuclear magnetic resonance (NMR) [13–15], the water in C–S–H gels was distinguished into three types: chemical bound water that is incorporated into the structure and form strong chemical bound with calcium silicate structure, physical bound water that is deeply adsorbed near the surface and capillary water without bound and diffuse freely in the capillary pore. Furthermore, quasi-elastic neutron scattering (QENS) technique [16] characterized the different water types by using the diffusion coefficient quantitatively. The unbound water molecules in the pore move in the fast rate (~10^{-9} m^2/s), while the water molecules confined in the gel pore diffuse slowly (~10^{-10} m^2/s). Similarly, hydration water dynamics in tri-calcium silicate (C$_3$S) pastes by QENS also showed that the diffusion coefficient of surface adsorbed water molecules decreases from 4×10^{-9} to 4×10^{-10} m^2/s after two days hydration. Currently, a proton field cycling relaxometry approach (PFCR) has been developed to investigate the dynamic properties of the water molecules bound in the C–S–H surface [17]. During approximately 1 ns, the diffusion coefficient of water molecules in the C–S–H gel pore surface is about 1/60 of the bulk water value (0.4 $\times$ 10^{-10} m^2/s).

2.2.4 Density and Water Content

The density of the C–S–H gel is closely related to the water content in the gel pore. The density for dried C–S–H has been measured by water pycnometry to be 2.85 g/cm^3 [18]. Recently, neutron and X-ray scattering techniques have been used to determine with great accuracy [7] the density and water content of the saturated globules. The density is 2.604 g/cm^3 and the chemical formula is C$_{1.7}$SH$_{1.8}$. These values include all water, evaporable and non-evaporable, within the saturated C–S–H particles, but do not include any adsorbed water on the surface or any other water outside of the particles. ^{1}H nuclear magnetic resonance (NMR) can fully characterize the nanoporosity of C–S–H in as-prepared material without the need for damaging sample drying. Density of C–S–H in sealed cured pastes changes as a function of the degree of hydration (α) and water to cement ratio. Densification of the C–S–H gel has been detected by the NMR technique. The C–S–H "solid" density, exclusive of gel pore water, slightly decreases from $\rho = 2.73$ g/cm^3 at $\alpha \approx 0.4$ to 2.65 g/cm^3 at $\alpha \approx 0.9$ due to an increase in the number of layers in the nanocrystalline aggregates. In the same range, the C–S–H "bulk" density, including gel water, increases from around 1.8 to 2.1 g/cm^3.

As shown in Fig. 2.6, Jennings [19] summarized the density evolution of the C–S–H at different water states. While the single sheet of C–S–H with all evaporable water removed has density 2.88 g/cm^3, a saturated globule without water adsorbed on the surface reduces to 2.602 g/cm^3. In section II, C–S–H gels with different water content are listed. As shown in Fig. 2.6a, fully saturated globule with monolayer on the surface and both interlayer and IGP full representing 11% RH. In Fig. 2.6b, partially dried with much of the interlayer water and monolayer removed. Volume of globule plus monolayer is reduced and density is increased from that of saturated globule. As shown in Fig. 2.6c, all evaporable water removed to represent the dried state. The IGP are empty leaving an internal void and reducing density. On rewetting water is returned to the IGP and a monolayer is present. The full IGP increases density (Fig. 2.6d).

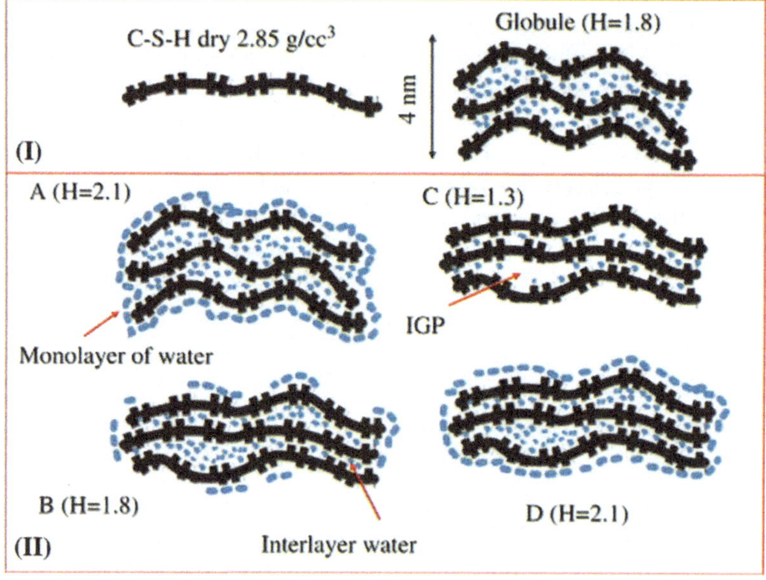

Fig. 2.6 A schematic of a globule with water contents representing various states achieved during drying from 11% RH and rewetting to various relative humidity and then returning to 11% RH [19]

2.2.5 Layered Feature

X-ray diffraction is a powerful tool to probe and understand the structure of nanocrystalline calcium-silicate-hydrates. Published XRD patterns from C–S–H of Ca/Si ratios ranging from ~0.6 to ~1.7 are fully compatible with nanocrystalline and turbostratic tobermorite [20] that is a layered calcium silicate crystal. From literature data, by studying the synthesized C–S–H gel, XRD patterns systematically have diffraction maxima at 7.4° (12.0 Å), 16.7° (5.3 Å), 29.1°(3.1 Å), 32.0° (2.8 Å), 49.8° (1.8 Å), 55.0° (1.7 Å) and 66.8° (1.4 Å). These maxima are broad and mostly asymmetric, and the maximum at 7.4° (12.0 Å) varies in position and intensity with C–S–H Ca/Si ratio [21–23]. Take the XRD intensity file from [22] for example, the patterns show a similarity to those of poorly crystalline 14 Å tobermorite, possessing broad basal reflections in the range 20–10 Å and further reflections at 5.30, 3.07 (strongest), 2.80, 1.83, 1.67, 1.53, 1.40, 1.19, 1.11, and 1.07 Å. Thus, it can be inferred that C–S–H has a layered nature. However, by comparing with other layered structures, line broadening is attributable to crystallite size in the nanometer range, whereas asymmetry is diagnostic of turbo-stratic disorder which is defined by the systematic occurrence, between successive layers, of random translations parallel to the layers (Fig. 2.7).

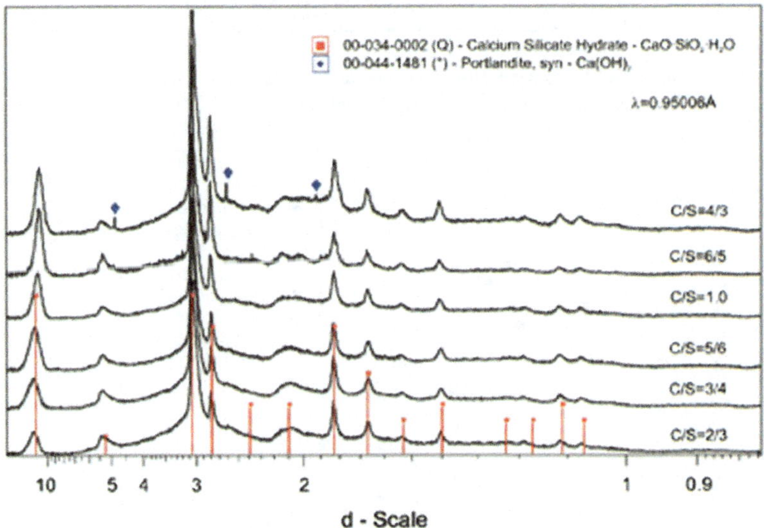

Fig. 2.7 Synchrotron-based X-ray diffraction (XRD) powder patterns of selected nanocrystalline calcium silicate hydrate (C–S–H) phases with different C/S ratios. Wavelength = 0.95006 Å [22]

Regarding the defective layered features of the C–S–H gel, the layered order cannot persist in long range. It can be observed from Fig. 2.8 that compared with the true tobermorite, the crystal length is dramatically reduced and the layered orientation is disturbed.

In addition, the composition can also influence the arrangement of the calcium silicate lamellas. The evolution of C–S–H XRD patterns as a function of the C–S–H Ca/Si ratio has been discussed Refs. [22, 23]. In these articles, the reflection between 13.5 and 11.2 Å which corresponds to the C–S–H *001* reflection, i.e. to the layer-to-layer distance [25], is sensitive to the Ca/Si ratio. Previous published XRD data are summarized and illustrated in Fig. 2.9 [21, 23, 26–33]. While at low Ca/Si ratio (<0.9), interlayer distance is close to 13.5 Å, with increasing Ca/Si ratio, this distance reduces down to 11.2 Å for Ca/Si = 1.7.

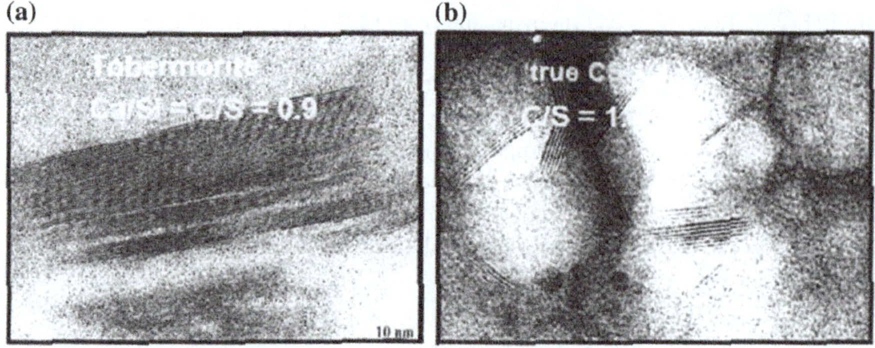

Fig. 2.8 TEM photos for the tobermorite and CSH. **a** A synthesized CSH with Ca/Si = 0.9, quite similar to the tobermorite structure. **b** The CSH gel with Ca/Si = 1.7 [24]

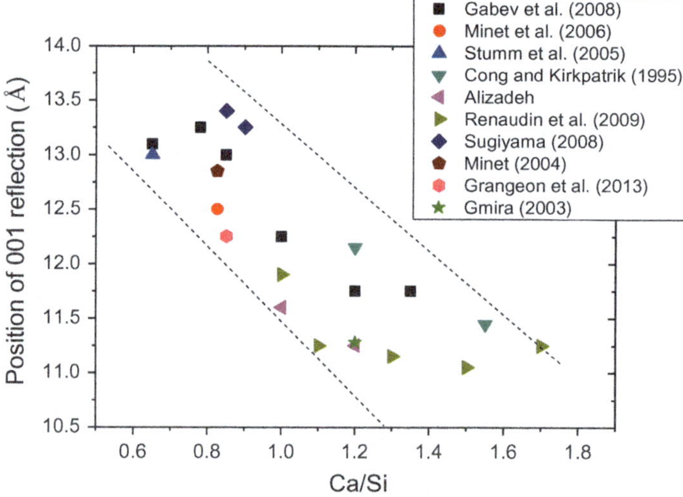

Fig. 2.9 Variation in the position of the C–S–H basal reflection as a function of structural Ca/Si ratio

2.2.6 *Silicate Polymerization*

As discussed in Sect. 2.2.5, it is widely accepted that the C–S–H has a layered structure akin mostly to tobermorite and jennite minerals. The major difference between the C–S–H gel and its structural analogues is the morphology of the silicate chain. The Nuclear magnetic resonance spectroscopy (NMR) is a powerful characterization method capable of providing information about the structure of materials. The method exploits the interaction of nonzero spin nuclei and an external magnetic field to gain information about the local atomic environment of a specific nucleus. From Si^{26} MAS NMR spectra, see the measurements in Fig. 2.10, it is known that the silicate chains in C–S–H are (i) linear (only Q_0, Q_1, and Q_2 peaks appear without traces of Q_3 and Q_4 peaks, where Q_0 accounts for non-hydrated clinker and Q_n means tetrahedrally coordinated Si atom with $n = 1, 2, 3$, and 4 bridging oxygens), and (ii) follow the Dreierketten arrangement (a Q_{2p} peak is resolved in the Q_2 peak, showing a silicate tetrahedron connected to two other tetrahedron). The defective calcium silicate layers result from the discontinuous silicate chains that are Dreierketten with finite lengths of $2, 5, 8, \ldots, 3n - 1$, where n is an integer. The arrangement of the silicate chains will be discussed in Sect. 2.3.

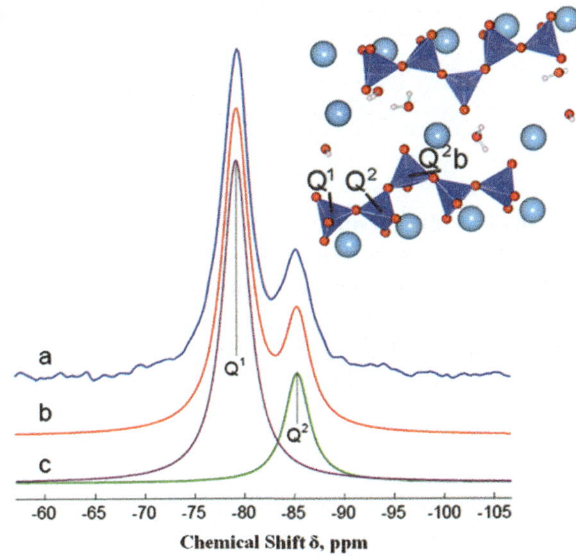

Fig. 2.10 ^{29}Si MAS NMR spectra of C–S–H gel. δ is the chemical shift in ppm. Note the decomposition according to Si position

As the Ca/Si ratio varies, the silicate structure changes significantly due to the interplay between silicate chains and counter ions Ca atoms. The major morphology variation is the polymerization of the silicate structure, which can be estimated by the mean silicate chain length from (NMR). The mean silicate chain length (number of silicate tetrahedron in a chain) is normally calculated from the Q_2/Q_1 integral intensity ratio:

$$\text{Mean Chain Length} = 2(1 + Q_2/Q_1)$$

where Q_1 and Q_2 are related to the population of end-chain and middle-chain silicate tetrahedron, respectively. As shown in Fig. 2.11, MCL obtained from NMR testing on various C–S–H samples [34–36], the silicate chain length progressively reduces with increasing Ca/Si ratio, despite some scattering. The depolymerization role of Ca atoms is an important characteristic in studying the calcium silicate skeleton.

Besides the Ca/Si ratio effect, polymerization of silicates is also attributed to the humidity level and temperature. The C–S–H in moist cured cement paste samples

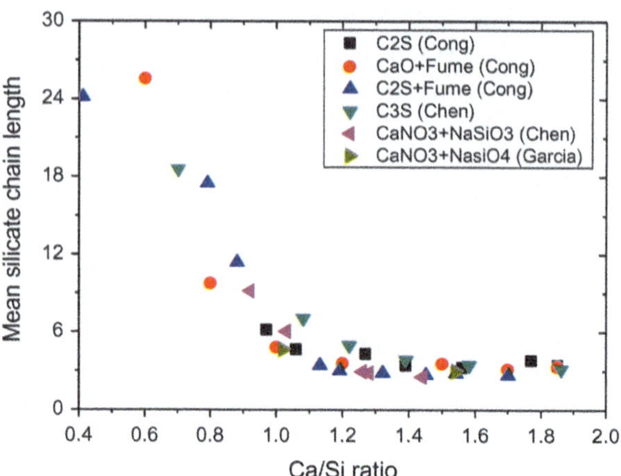

Fig. 2.11 Mean silicate chain evolution with Ca/Si ratio

virtually contains all dimers, whereas pastes cured at the low relative humidity conditions (below 10% RH) have a greater degree of polymerization [37]. The ^{29}Si MAS NMR was employed to investigate the nanostructural changes in the silicate structure of the C–S–H preparations dried at various increments from 11% RH condition [38]. It is observed in Fig. 2.12 that the extent of silicate polymerization increases at higher mass loss levels more noticeably after a structural water loss of about 5% in mass.

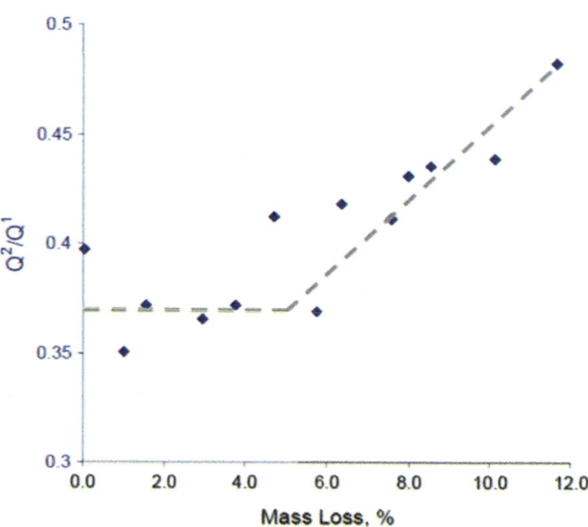

Fig. 2.12 Increase of the C–S–H polymerization on the incremental removal of water from 11% RH conditioned powder

2.2.7 Mechanical Properties

As shown in Fig. 2.13a, nanoindentation consists of making contact between a sample and an indenter tip of known geometry and mechanical properties, followed by a continuously applied and recorded change in load, P, and depth, h, which can offer two indentation mechanical properties: indentation modulus (M) and hardness (H) of nanoscale clusters of randomly oriented C–S–H particles at micrometer scale. The recent nanoindentation test shows that a low stiffness C–S–H gel with indentation modulus 25 ± 2 GPa and a high stiffness phase with indentation modulus 30 ± 5 GPa [6]. The discrepancy difference of stiffness is attributed to the particles packing density. By using the micro-mechanics scaling relation such as self-consistent scheme (SC) and MT model, the relationship between indentation hardness and porosity has been constructed, which obtain the indentation modulus of C–S–H gel around 65 GPa and the hardness around 4 GPa.

2.3 Mineral Analogues of C–S–H Gel

There are at least thirty crystalline minerals [8] that resemble the composition of C–S–H gel. According to the TEM [40] and XRD investigations [28], C–S–H gel exhibits layered features of the tobermorite [20] and jennite structure [41].

2.3.1 Tobermorite

Tobermorite is a calcium silicate hydrate mineral that exists in nature. Three different types of tobermorite exist, i.e., 14 Å, 11 Å or 9 Å tobermorite. They are distinguished on the basis of different basal spacing corresponding to the differed hydration degree [42]. The 11 Å tobermorite has a layer structure and consists of a central sheet of calcium with silicate chains sandwiched. The so-called Dreierketten chains are shown in Fig. 2.14 that two "paired" tetrahedrons pointed toward Ca sheet and one "bridging" tetrahedron linked with interlayer form a basic unit (period) for repetition. The nearest silicate chains are not aligned "head to head" but rather shifted by $b/2$. Water molecules and calcium atoms can be located in the interlayer zones. As confirmed by the NMR measurements, the Q_2 silicate sites are dominant, implying the infinite length of the silicate chain. Additionally, tobermorite 11 Å has two different structures: Hamid structure which is a refinement that describes the tobermorite as independent layers [20] and Merlino structure which considers tobermorite as chemically bonded layers [43]. While the calcium silicate backbone layer is structurally unchanged, Hamid tobermorite can have three different Ca/Si ratios, namely 0.67, 0.83 and 1 with chemical formula $Ca_4 [Si_6O_{14}(OH)_4]\cdot 2H_2O$, $Ca_5 [Si_6O_{16}(OH)_2]\cdot 2H_2O$ and $Ca_6 [Si_6O_{18}]\cdot 2H_2O$. The Ca/Si ratio variation is caused by adding the calcium atoms in the interlayer region; meanwhile, the protons in the hydroxyl groups should be removed to maintain the electronic neutrality. The

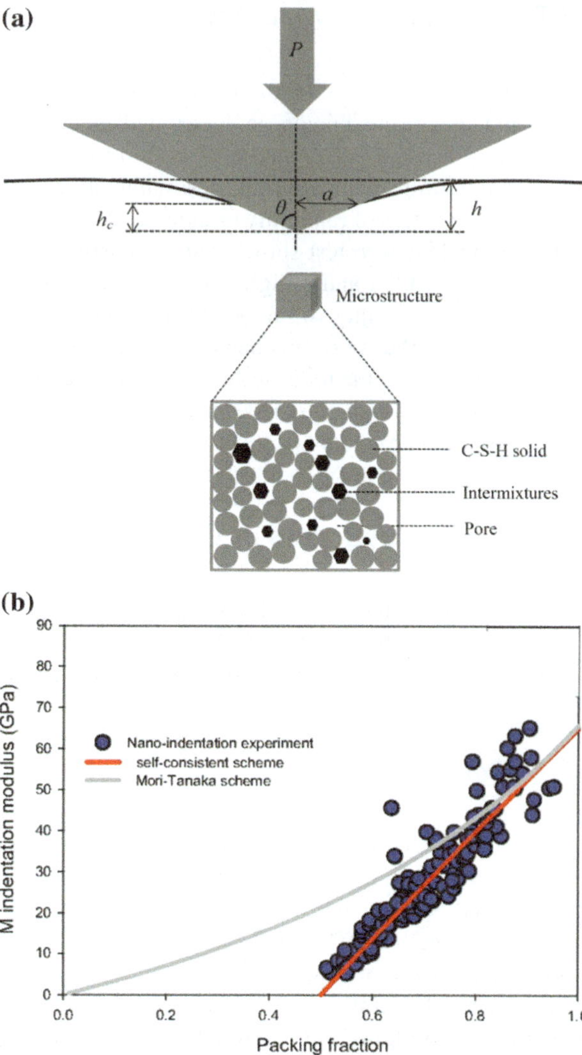

Fig. 2.13 **a** Geometry of a nanoindentation test on the "C–S–H" phase; **b** relationship between packing density and indentation modulus [39]

major difference between Hamid and Merlino structures is the interlayer connection between the bridging tetrahedrons. As shown in Fig. 2.14d and e, the head-to-head connection of bridging tetrahedrons forms an ionic-covalent Si–O–Si bond with the neighboring calcium silicate sheets. While Q_2 species changes to Q_3 species, the 2D single silicate chains transforms to the 3D branch and network. Furthermore, thermal behavior of tobermorite investigated by multi experimental techniques indicates the transformation from tobermorite 14 to 9 Å: tobermorite 14 Å transforms into tobermorite 11 Å by heating up to 80–100 °C and further heating up to 300 °C for transformation to 9 Å.

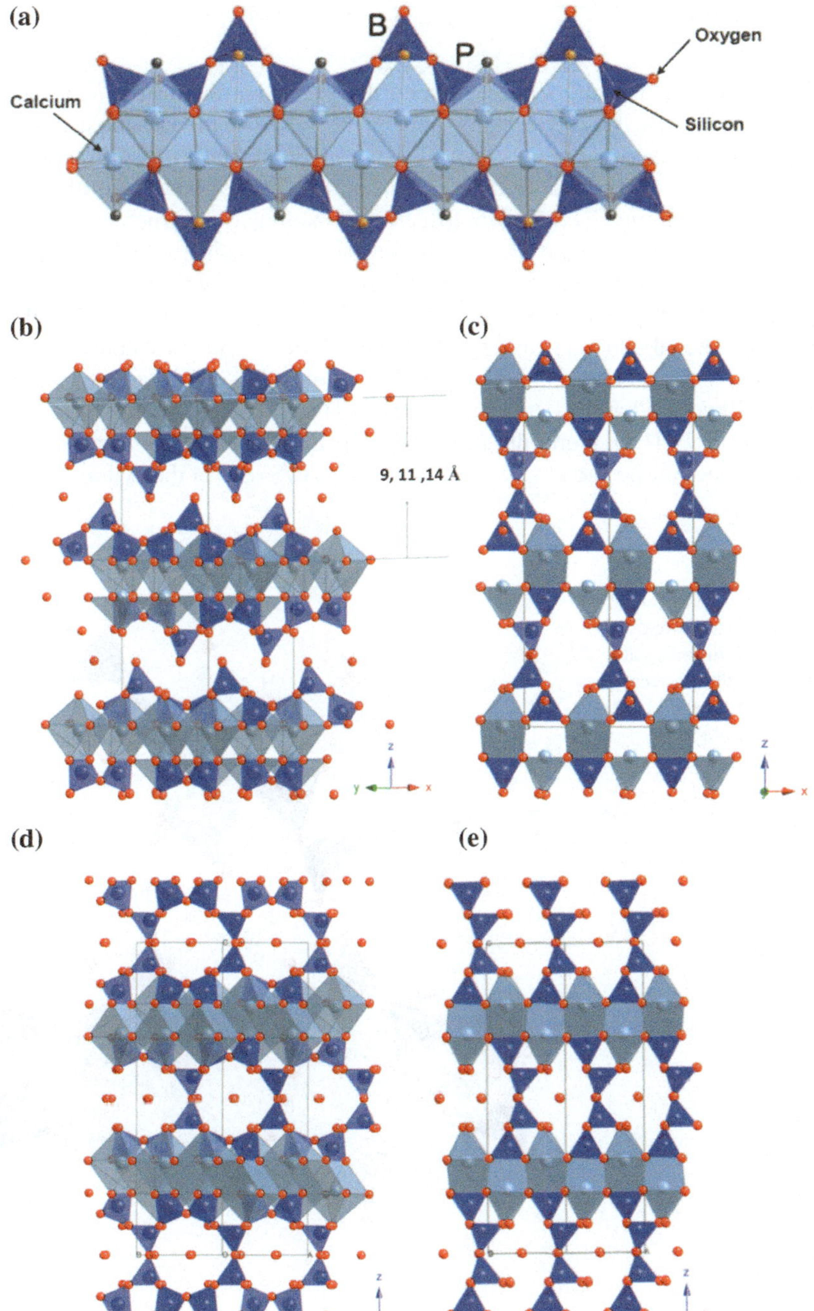

Fig. 2.14 a A schematic of C–S–H layer showing a main Ca–O layer with silicate tetrahedron attached on both sides. Crystal structure of tobermorite 11 Å projected along **b** [110] and **c** [010] proposed by Hamid, **d** [110] and **e** [010] by Merlino. Silicate chains, calcium octahedral, and oxygen atoms are shown as dark blue, light blue tetrahedron, and red spheres, respectively

2.3.2　*Jennite*

Jennite is a natural and rare silicate mineral that has been found in several regions in the world such as Israel, Germany, and Japan. In 2004, the crystal structure of jennite has been solved by Bonaccorsi et al. using single-crystal X-ray diffraction data collected on a very thin crystal from Fuka, Japan [41]. The solved crystal structure of jennite is shown in Fig. 2.15, in two different projections. The structure consists of three distinct modules: ribbons of edge-sharing calcium octahedral, silicate chains of wollastonite-type, and additional calcium octahedron in special position on inversion centers. The ribbons are connected to each other to give a zigzag layer of calcium octahedron. The ribbons are also firmly linked through silicate chains. From the analysis of hydrogen bonds connecting the H_2O molecules, it is concluded that both oxygen and hydroxyl anions coexist in jennite. The chemical formula of jennite is $Ca_9Si_6O_{18}(OH)_6 \cdot 8H_2O$. Here, the C/S ratio in jennite reaches 1.5, a value greater than that in tobermorite and much closer to the mean C/S ratio of the hydrated cement in reality.

Fig. 2.15 Dreierkette chains present in the structure of jennite (which in theory are of infinite length) projected along [010] (top) and [100] (bottom). "Paired" and "bridging" tetrahedron are labeled "P" and "B", respectively

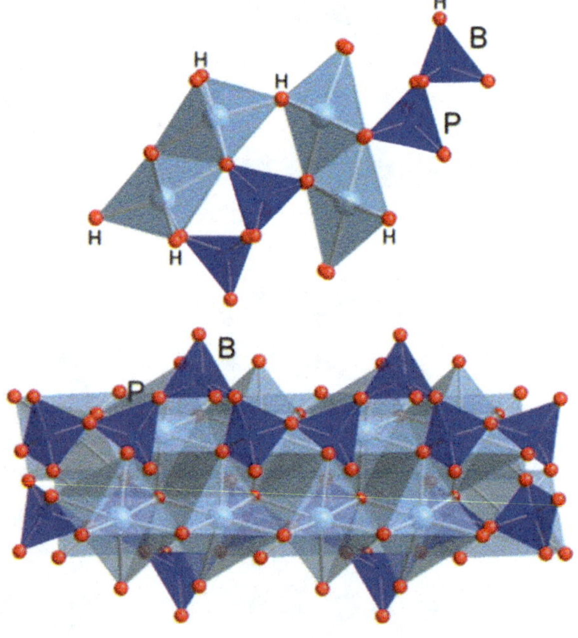

2.4 Models of the C–S–H Gel

2.4.1 Models for the Nanostructure and Morphology

Several microstructural models have been proposed to describe the structure of
C–S–H gel at nanoscale and formation of gel pores during the past years. Of the
several models, four models, proposed by Powers and Brownyard, Feldman and
Sereda, Wittmann, and Jennings and Tennis, have been utilized extensively [44].

2.4.1.1 Powers and Brownyard Model

Powers and Brownyard [45] proposed a model for the structure of cement gel based
on their comprehensive studies of water vapor sorption isotherms and chemically
bound water in hardening cement pastes. In their model, the C–S–H gel is composed
of particles that have a layered structure, which are composed of 2–3 sheets. As
shown in Fig. 2.16, the randomly distributed layers are bonded together by surface
forces with strong ionic-covalent bonds linking adjacent particles.

Based on their studies, water molecules in the C–S–H gel are categorized into
three types: water of constitution, gel water, and capillary water. Water of constitution involves water of crystallization and chemically bound, non-evaporable
water, including the hydroxyl groups associated with metallic ions Ca–OH. Gel
water includes adsorbed water and water bound by physical surface forces. These
water molecules are packed between the layers of the crystal hydrate. On the other

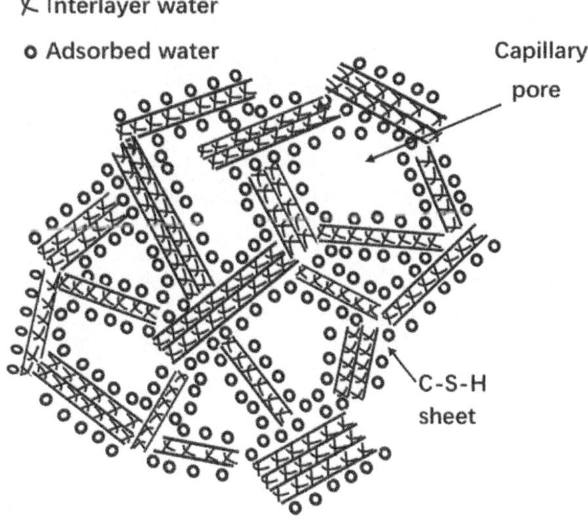

Fig. 2.16 Schematic diagram of the Powers and Brownyard model

hand, the capillary water is free water insides the pores, not bound by the surface force.

2.4.1.2   Feldman and Sereda Model

Feldman and Sereda proposed a model similar to the Powers–Brownyard model with some important differences [46]. As shown in Fig. 2.17, the sheets composing the C–S–H gel have an irregular array of single layers, two to four molecules thick, which may come together randomly to create interlayer space. Based on gas sorption isotherm results, they concluded that water enters vacated interlayer spaces of C–S–H at low relative humidity and is structurally incorporated in the C–S–H structure. The interlayer water can reversibly enter and leave the interlayer spaces, a feature that challenged the model by Powers and Brownyard and questioned the interpretation of adsorption data experiments. Bonding between layers is considered to be through solid–solid contacts, which are considered as connections intermediate in feature between weak van der Waals and strong ionic-covalent bonds. The solid–solid contacts between the layers forming during drying of the cement paste and are disrupted by wetting. Interlayer bonding is a special kind of chemical bonding and cannot be considered as resulting from interactions between free surfaces of C–S–H layers.

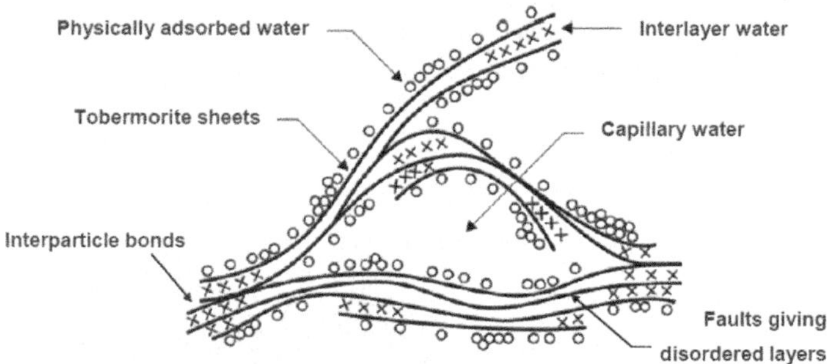

Fig. 2.17   Schematic illustration of the Feldman and Sereda model

Because adsorption of water molecules within the interlayer region of the C–S–H gel distorts the measurement due to water vapor, they believed that nitrogen adsorption measuring surface area of pores is more closely than the water adsorption method.

2.4.1.3 Munich Model

The Munich model developed by F. Wittmann, 1979, was proposed based on sorption measurements and conceived the structure of C–S–H gel as a threedimensional network of amorphous colloid gel particles forming a xerogel in Fig. 2.18 [47]. The C–S–H gel particles are separated by strongly adsorbed thin films of water that adopt a repulsive swelling pressure to counteract the attractive van der Waals interactions between the solid surfaces. Van der Waals forces contribute to the binding energy between gel particles, but strong ionic-covalent bonds still predominate. The Munich model suggests that the disjoining pressure of the water in a capillary pore could cause the C–S–H gel to swell or shrink as a function of relative humidity in the pressure ranges associated with capillary condensation. The Munich model has been used to explain the behavior of hardened cement paste under different humidity conditions.

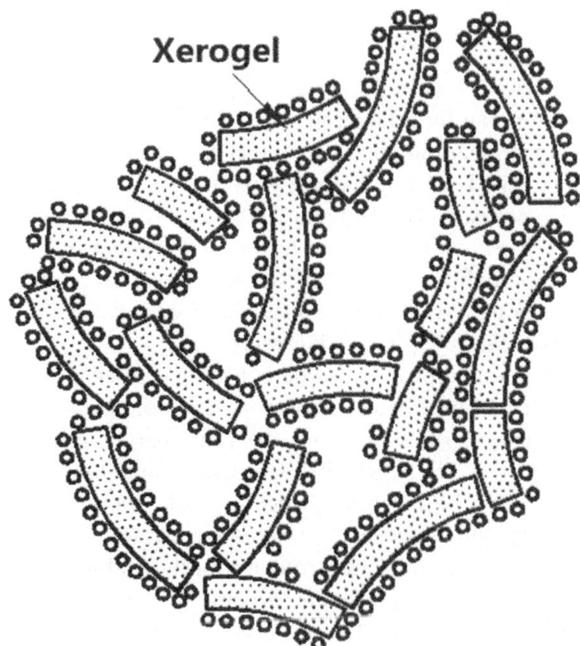

Fig. 2.18 Schematic illustration of Wittmann's model

2.4.1.4 Jennings Model

Compared with previous models, the Jennings model gives a comprehensive description for the C–S–H gel in the respects of the surface area, density, and water content [19, 48]. In the model, Jennings proposed that the C–S–H gels are an aggregation of the colloids with nanosize. The smallest building blocks are colloid spheres with

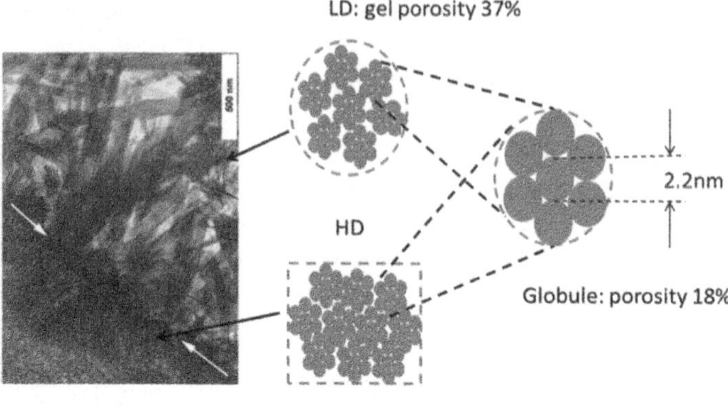

Fig. 2.19 Schematic representation for the CM-I

diameter 2.2 nm and density 2.8 g/cm^3. As shown in Fig. 2.19, agglomeration of these spheres can form larger globules with diameter 5.6 nm. Those globules aggregated together according to different packing efficiency can result in two main species of the C–S–H gels: high density (HD) with 24% porosity and low density (LD) with 37% porosity. The HD and LD phases of CSH gels can explain the morphologies of dense inner product and loose outer products respectively. This can help interpret the contradictory results of surface area from multi techniques. For example, even though the HD phase occupies large surface area/volume, the N$_2$ cannot penetrate into the narrow pores in HD phases.

These two phases also demonstrate discrepancy in mechanical properties. Later Ulm performed extensive nanoindentation tests on different cement paste and the bimodal distribution of the elastic properties proved the presence of LD and HD phases of C–S–H gel [49]. But Ulm pointed that the packing efficiencies for two phases are 74 and 64%, which are corresponding to the maximum for the sphere packing and closed-packed random packing respectively.

To improve the inner structure for the building block, Jennings took into consideration the layer molecular structure of tobermorite and Jennite and modified the basic building units. The density of globule (2.604 g/cm^3) and the average chemical composition of C–S–H (C$_{1.7}$SH$_{1.8}$) were achieved from the SANS and SAXS tests [7]. As shown in Fig. 2.20, the C–S–H is treated as an assembly of brick-shaped globules with cross section of 5 nm rather than the sphere colloid. According to different humility, the C–S–H globule can be distinguished into four states: the fully saturated globule with monolayer; partial saturated globule; saturated without monolayer water; dried state with evaporated water removed. These globules pack together and between the globules exist different spaces: the intra-globule spaces (IGP), the small gel pores (SGP), and the large gel pores (LGP), which have been discussed in section water and C–S–H density.

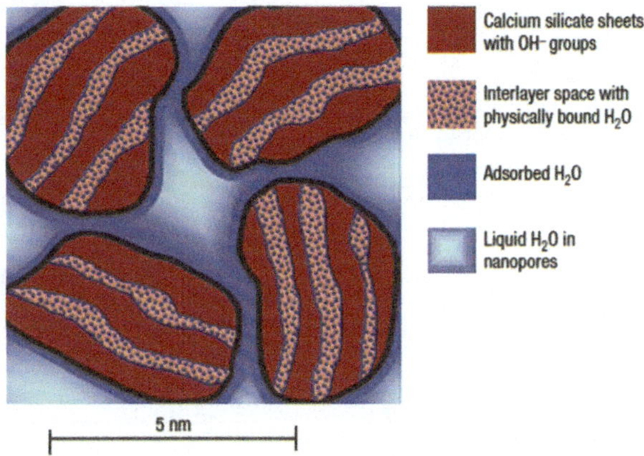

Fig. 2.20 Schematic diagram for the CM-II model [7]

2.4.2 Models for the Atomic Structure

The molecular structure of the C–S–H gel has been investigated by many researchers in the physics and chemistry fields. As discussed in Sect. 2.3, the C–S–H gel are characterized by the layered features, tobermorite analogues and silicate chains distribution that are predicted from many experiments such as the XRD, NMR and TEM [29, 35, 40].

Bernal et al. proposed the first dreierkette-based model for C–S–H [50]. By means of crystallographic investigation, they suggested that the hydration product of C_3S is similar to the C–S–H formed in dilute suspension, i.e. a layered fibrous structure associated with 1.1 nm tobermorite. The suggestion by Taylor and Howison [51] that the generally accepted Ca/Si ratio of about 0.8 for tobermorite can be increased by omitting the bridging tetrahedron (replaced by calcium atoms), which is one of the features in many other tobermorite-based models for C–S–H [36, 52–55]. In order to explain the variety of compositions, it was suggested that the C–S–H layers are separated by the so-called interlayer region, in which H_2O, Ca^{2+} and OH^- can be present. A few other studies proposed that the C–S–H has a structure related to that of jennite [56]. In jennite, half of the oxygen atoms come from central Ca–O part of the C–S–H layer which shares with OH groups, whereas in tobermorite all the oxygen atoms are shared with the infinite parallel rows of silicate chains.

Taylor proposed that C–S–H in cement paste is evidently a disordered layer structure and is a mixture of structures based on both 1.4 nm tobermorite and jennite [57]. As shown in Fig. 2.21, this was theoretically one of the possible ways to explain the high Ca/Si ratio of C–S–H in cement paste (1.7–2.0) as dimeric 1.4 nm tobermorite and jennite type structures could have a C/S ratio of about 1.25 and 2.25 at early ages, respectively. Moreover, the water to calcium ratio of the C–S–H at 11% RH lies between that of the suggested structural representatives.

Currently, Richardson and Groves [58] have proposed a twofold classification to clarify C–S–H chemistry, which references tobermorite/jennite (T/J) models and

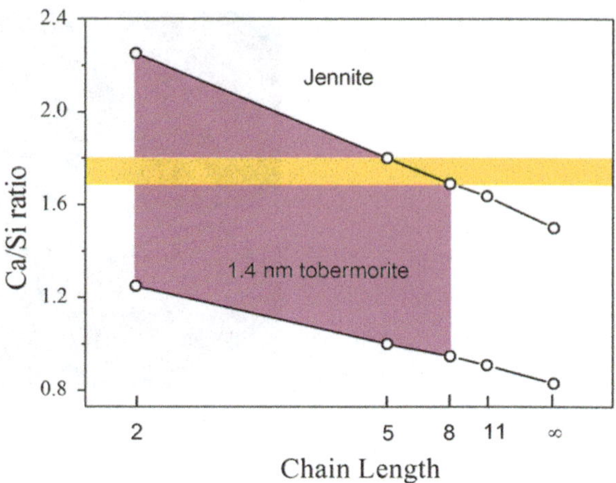

Fig. 2.21 Relations between Ca/Si and silicate chain length in terms of tobermorite and jennite based on tobermorite and jennite structures [57]

tobermorite/calcium hydroxyl (T/CH) models. While the T/CH class considers models that are solid solutions of tobermorite layers sandwiching calcium hydroxide, the T/J class considers C–S–H as an assembly of tobermorite regions followed by jennite domains. The distinction between main layer and interlayer Ca^{2+} and the consideration of water content of C–S–H at various drying conditions were considered lacking by Taylor. However, the T/CH class has been found to be relevant for hydrated KOH-activated metakaolin Portland cement, more common water activated Portland cement pastes can be only partly described by the T/J or T/CH approaches.

2.4.3   Models Based on Molecular Simulation

Even though the modified tobermorite and jennite models discussed above demonstrate some similarity with the C–S–H gel, these C–S–H analogues do not possess strict structural and morphological resemblance to those in realistic C–S–H. These theoretical models have not comprehensively considered the basic physical properties such as the density, Ca/Si ratios, silicate chain distribution and water content, and so on. For example, Ca/Si ratio and density values show discrepancy between mineral analogues (tobermorite: Ca/Si = 0.83, 2.18 g/cm^3) and the C–S–H gel (Ca/Si = 1.75, 2.6 g/cm^3).

Based on the perspective of the chemical composition of C–S–H as the essential properties, the "realistic model" of C–S–H gel was constructed by the assistance of the molecular simulation [39]. The consistent molecular model, systematically considering the properties of C–S–H gel obtained from multi experiments, has demonstrated the interaction between CaO, SiO_2, and H_2O to form a defected silica chain morphology that links to realistic values for density and chemical composition of an averaged C–S–H phase. As shown in Fig. 2.22, the model, with a chemical compo-

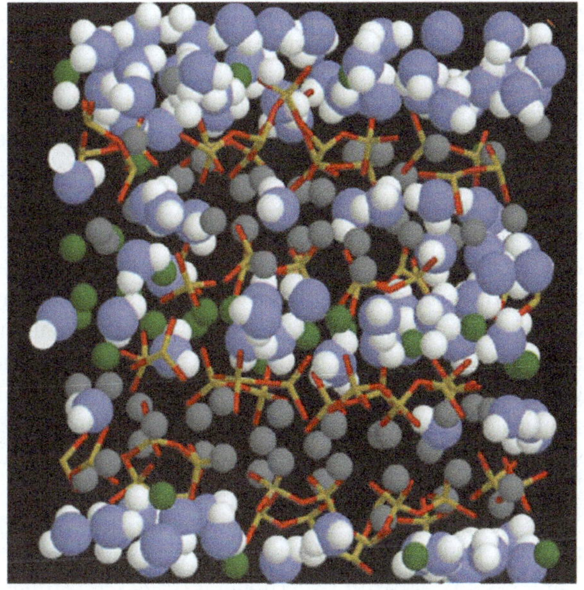

Fig. 2.22 The molecular model of C–S–H: the blue and white spheres are oxygen and hydrogen atoms of water molecules, respectively; the green and gray spheres are inter and intra-layer calcium ions, respectively; yellow and red sticks are silicon and oxygen atoms

sition of $(CaO)_{1.65}(SiO_2)(H_2O)_{1.75}$, suggests that the C–S–H gel structure includes both glass-like short-range order and crystalline features of the mineral tobermorite. Validated by experiments, the model also predicts several essential structural features and fundamental properties such as stiffness and strength of real cement paste system.

Realistic model is the first attempt of decoding of the cement structure at nanoscale by molecular simulation, which can bridge the experimental findings and theoretical models. Followed by the "realistic model", thermodynamics, dynamics, reactivity, and mechanical properties have been extensively investigated, which brings about molecular insights and bridges the gap between microscale structure and macro-level laboratory findings [59–61]. However, the model still has some limitation: the silicate chain backbone, hydroxyl groups and the local structure of calcium octahedron should be further improved to accurately describe the molecular structure of C–S–H gel. These deficiencies of the "realistic model" will be discussed in detail in Chap. 4. On the other hand, complicated composition, including Ca/Si ratio and water contents, has not been considered sufficiently in the realistic model.

2.5 Chapter Summary

The properties of the C–S–H gel have been investigated by multi experimental techniques, which provide molecular insights on the density, Ca/Si ratio, water states, and the silicate chain distributions. The useful experimental information offers the vague first expression about the C–S–H gel: the disordered glassy layers that in some extent resemble the mineral analogue tobermorite or jennite, the density at saturated state around $2.6\ g/cm^3$, the defective silicate chains with Q_1 species occupying dominated percentage, various Ca/Si ratios with mean value at 1.75 and the intrinsic indentation modulus is around 65 GPa. These key features of the C–S–H gel obtained from different techniques should be combined together to form a model for completely understanding the C–S–H gel. It is accepted that due to the compositional diversity and hierarchy nature of the C–S–H gel, only experimental techniques alone are restricted with instrumental accuracy at the relevant length scale and time scale, and the materials' purity. Hence, various theoretical models have been constructed at colloid level and molecular level to explain the experimental finding. While many crystal minerals such as tobermorite, jennite, portlandite, or their disordered combinations are proposed as models to explain various structural features of the C–S–H gel, the structure of C–S–H gel itself has not been fully understood yet, especially at nanoscale. Major challenge preventing the theoretical investigation is the inconsistence in the Ca/Si ratio, the silicate chain length, the density, and the disordered layered structure compared to tobermorite.

Molecular simulation acts as a bridge between theory and experiment. The theory may be tested by conducting a simulation using the same model. The simulated model can be compared with experimental results to estimate the value of the theoretical models. On the other hand, because the intrinsic molecular structure of the C–S–H gel cannot be directly accessed by experimental exploring, computational methods can help to interpret the experimental results and play a valuable role in understanding the structural and dynamic properties at the molecular level. In the next chapter, various molecular simulation methods will be introduced, which further are employed in the model construction and application in Chaps. 4–7.

References

1. Zhang, L. (2012). *The structure study of calcium silicate hydrate (C–S–H) (QE report)*. Hong Kong University of Science and Technology.
2. Taylor, H. F. W. (1997). *Cement chemistry*. London: Academic Press.
3. Bishop, M., Bott, S. G., & Barron, A. R. (2003). A new mechanism for cement hydration inhibition: Solid-state chemistry of calcium nitrilotris(methylene)triphosphonate. *Chemistry of Materials, 15*(16), 3074–3088.
4. Groves, G. W. (1986). TEM studies of cement hydration. *MRS Online Proceeding Library Archive, 85*.
5. Nomat, A. (2004). The structure and stoichiometry of C–S–H. *Cement and Concrete Research, 34*, 1521–1528.
6. Costantinide, G., & Ulm, F. (2006). The nanogranular nature of C–S–H. *Journal of Mechanics and Physics of Solids, 55*, 64–90.
7. Allen, A. J., Thomas, J. J., & Jennings, H. M. (2007). Composition and density of nanoscale calcium silicate hydrate in cement. *Nature Material, 6*, 311–316.
8. Richardson, I. J. (1999). The nature of C–S–H in hardened cements. *Cement and Concrete Research, 29*(8), 1131–1147.
9. Richardson, I. G., & Groves, G. W. (1993). Microstructure and microanalysis of hardened ordinary Portland cement pastes. *Journal of Material Science, 28*, 265–277.

10. Hou, D., Ma, H., Zhu, Y., & Li, Z. (2014). Calcium silicate hydrate from dry to saturated state: Structure, dynamics and mechanical properties. *Acta Materialia, 67,* 81–94.
11. Hou, D. S., & Li, Z. J. (2014). Molecular dynamics study of water and ions transported during the nanopore calcium silicate phase: Case study of jennite. *Journal of Materials in Civil Engineering, 26*(5).
12. Ma, H., & Li, Z. (2013). Realistic pore structure of Portland cement paste: Experimental study and numerical simulation. *Computer and Concrete, 11*(4), 317–336.
13. Wang, P. S., Ferguson, M. M., Eng, G., Bentz, D. P., Ferraris, C. F., & Clifton, J. R. (1998). 1H nuclear magnetic resonance characterization of Portland cement: Molecular diffusion of water studied by spin relaxation and relaxation time-weighted imaging. *Journal of Material Science, 33,* 3065–3071.
14. Rakiewicz, E. F., Benesi, A. J., Grutzeck, M. W., & Kwan, S. (1998). Determination of the state of water in hydrated cement phases using deuterium NMR spectroscopy. *Journal of the American Chemical Society, 120*(25), 6415–6416.
15. Greener, J., Peemoeller, H., Choi, C., Holly, R., Reardon, E. J., Hansson, C. M., et al. (2000). Monitoring of hydration of white cement paste with proton NMR spin–spin relaxation. *Journal of the American Ceramic Society, 83*(3), 623–627.
16. Bordallo, H. N., Aldridge, L. P., & Desmedt, A. (2006). Water dynamics in hardened ordinary Portland cement paste or concrete: From quasielastic neutron scattering. *Journal of Physics and Chemistry B, 110,* 17966–17976.
17. Korb, J. P., Monteilhet, L., McDonald, P. J., & Mitchell, J. (2007). Microstructure and texture of hydrated cement-based materials: A proton field cycling relaxometry approach. *Cement and Concrete Research, 37*(3), 295–302.
18. Brunauer, S., Kantro, D. L., & Copeland, L. E. (1958). The stoichiometry of the hydration of β-dicalcium silicate and tricalcium silicate at room temperature. *Journal of the American Chemical Society, 80*(4), 761–767.
19. Jennings, H. M. (2008). Refinements to colloid model of C–S–H in cement: CM II. *Cement and Concrete Research, 38*(3), 275–289.
20. Hamid, S. (1981). The crystal structure of the 11 A natural tobermorite $Ca_{2.25}Si_3O_{7.5}(OH)_{1.5} \cdot H_2O$. *Zeitschrift fur Kristallographie, 154,* 189–198.
21. Garbev, K., Bornefeld, M., Beuchle, G., & Stemmermann, P. (2008). Cell dimensions and composition of nanocrystalline calcium silicate hydrate solid solutions. Part 2: X-ray and thermogravimetry study. *Journal of the American Ceramic Society, 91*(9), 3015–3023.
22. Garbev, K., Beuchle, G., Bornefeld, M., Black, L., & Stemmermann, P. (2008). Cell dimensions and composition of nanocrystalline calcium silicate hydrate solid solutions. Part 1: Synchrotron-based X-ray diffraction. *Journal of the American Ceramic Society, 91*(9), 3005–3014.
23. Renaudin, G., Russias, J., Leroux, F., Cau-dit-Coumes, C., & Frizon, F. (2009). Structural characterization of C–S–H and C–A–S–H samples—Part II: Local environment investigated by spectroscopic analyses. *Journal of Solid State Chemistry, 182*(12), 3320–3329.
24. Gmira, A. (2003). Etude texturale et thermodynamique d'hydrates modèles du ciment. Orléans.
25. Grangeon, S., Claret, F., Lerouge, C., Warmont, F., Sato, T., Anraku, S., et al. (2013). On the nature of structural disorder in calcium silicate hydrates with a calcium/silicon ratio similar to tobermorite. *Cement and Concrete Research, 52,* 31–37.
26. Alizadeh, R., Raki, L., Makar, J. M., Beaudoin, J. J., & Moudrakovski, I. (2009). Hydration of tricalcium silicate in the presence of synthetic calcium–silicate–hydrate. *Journal of Materials Chemistry, 19*(42), 7937–7946.
27. Cong, X., & Kirkpatrick, R. J. (1995). Effects of the temperature and relative humidity on the structure of C–S–H gel. *Cement and Concrete Research, 25*(6), 1237–1245.
28. Gmira, A. (2003). Etude textural et thermodynamique d'hydrates modeles du ciment. Ph.D. thesis, Universite D Orieans, France.
29. Grangeon, S., Claret, F., Linard, Y., & Chiaberge, C. (2013). *X-ray diffraction: A powerful tool to probe and understand the structure of nanocrystalline calcium silicate hydrates.* Acta Crystallographica Section B: Structural Science Crystal Engineering.

30. Minet, J., Abramson, S., Bresson, B., Sanchez, C., Montouillout, V., & Lequeux, N. (2004). New layered calcium organosilicate hybrids with covalently linked organic functionalities. *Chemistry of Materials, 16*(20), 3955–3962.

31. Minet, J., Abramson, S., Bresson, B., Franceschini, A., Van Damme, H., & Lequeux, N. (2006). Organic calcium silicate hydrate hybrids: A new approach to cement based nanocomposites. *Journal of Materials Chemistry, 16*(14), 1379–1383.

32. Stumm, A., Garbev, K., Beuchle, G., Black, L., Stemmermann, P., & Nüesch, R. (2005). Incorporation of zinc into calcium silicate hydrates. Part I: Formation of CSH (I) with C/S = 2/3 and its isochemical counterpart gyrolite. *Cement and Concrete Research, 35*(9), 1665–1675.

33. Sugiyama, T., Ritthichauy, W., & Tsuji, Y. (2008). Experimental investigation and numerical modeling of chloride penetration and calcium dissolution in saturated concrete. *Cement and Concrete Research, 38*(1), 49–67.

34. García-Lodeiro, I., Fernández-Jiménez, A., Sobrados, I., Sanz, J., & Palomo, A. (2012). C–S–H gels: Interpretation of 29Si MAS-NMR Spectra. *Journal of the American Ceramic Society, 95*(4), 1551–2916.

35. Chen, J. J., Thomas, J. J., Taylor, H. F. W., & Jennings, H. M. (2004). Solubility and structure of calcium silicate hydrate. *Cement and Concrete Research, 34*(9), 1499–1519.

36. Cong, X., & Kirkpatrick, R. (1996). 29Si MAS NMR study of the structure of calcium silicate hydrate. *Advanced Cement Based Material, 3*(3–4), 144–156.

37. Macphee, D. E., Lachowski, E. E., & Glasser, F. P. (1998). Polymerization effects in C–S–H: Implications for Portland cement hydration. *Advances in Cement Research, 1*(3), 131–137.

38. Alizadeh, R. A. (2009). Nanostructure and engineering properties of basic and modified calcium silicate hydrate systems. Ph.D. thesis of University of Ottawa.

39. Pellenq, R. J. M., Kushima, A., Shahsavari, R., Van Vliet, K. J., Buehler, M. J., & Yip, S. (2009). A realistic molecular model of cement hydrates. *PNAS, 106*(38), 16102–16107.

40. Groves, G. (1987). TEM studies of cement hydration. *Materials Research Society Symposium Proceedings, 85,* 3–12.

41. Bonaccorsi, E., Merlino, S., & Taylor, H. F. W. (2004). The crystal structure of Jennite $Ca_9Si_6O_{18}(OH)_6 \cdot 8H_2O$. *Cement and Concrete Research, 34*(9), 1481–1488.

42. Shahsavari, R., Buechler, M. J., Pellenq, R. J. M., & Ulm, F. J. (2009). First-principles study of elastic constants and interlayer interactions of complex hydrated oxides: Case study of tobermorite and jennite. *Journal of American Ceramic Society, 92*(10), 2323–2330.

43. Merlino, S., Bonnacorsi, E., & Armbruster, T. (2001). The real structure of tobermorite 11 A: Normal and anomalous forms, OD character and polytypic modifications. *European Journal of Mineralogy, 13*(3), 577–590.

44. Aligizaki, K. K. (2006). *Pore structure of cement-based materials: Testing, interpretation and requirements.* CRC Press.

45. Powers, T. C., & Brownyard, L. (1946). Studies of the physical properties of hardened Portland cement paste. *ACI Journal Proceedings, 43.*

46. Feldman, R. F., & Sereda, J. P. (1968). A model for hydrated Portland cement paste as deduced from sorption-length change and mechanical properties. *Matériaux et Construction, 1*(6), 509–520.

47. Wittmann, F. H. (1979). Trends in research on creep and shrinkage of concrete. *Cement Production and Use,* 143–161.

48. Jennings, H. (2000). A model for the microstructure of calcium silicate hydrate in cement paste. *Cement and Concrete Research, 30,* 101–116.

49. Costantinides, G., & Ulm, F. (2004). The effect of two types of C–S–H on the elasticity of cement-based materials: Result from nanoindentation and micromechanical modeling. *Cement and Concrete Research, 34,* 67–80.

50. Bernal, J. D., Jeffery, J. W., & Taylor, H. F. W. (1952). Crystallographic research on the hydration of Portland cement. A first report on investigations in progress. *Magazine of Concrete Research, 4*(11), 49–54.

51. Taylor, H. F. W., & Howison, J. W. (1956). Relationships between calcium silicates and clay minerals. *Clay Minerals Bulletin, 3,* 98–111.

52. Kantro, D. L., Brunauer, S., & Weise, C. H. (1962). Development of surface in the hydration of calcium silicates. II. Extension of investigations to earlier and later stages of hydration. *The Journal of Physical Chemistry, 66*(10), 1804–1809.
53. Stade, H. (1980). On the structure of ill-crystallized calcium hydrogen silicates. 2. A phase consisting of polysilicate and disilicate. *Zeitschrift fur Anorganische und Allgemeine Chemie, 470*(11), 69–83.
54. Cong, X., & Kirkpatrick, R. J. (1996). 29Si and 17O NMR investigation of the structure of some crystalline calcium silicate hydrate. *Advances in Cement Based Materials, 3*(3), 133–143.
55. Grutzeck, M. W. (1999). A new model for the formation of calcium silicate hydrate (CSH). *Material Research Innovations, 3*(3), 160–170.
56. Viehland, D., Yuan, L. J., Xu, Z., Cong, X. D., & Kirkpatrick, R. J. (1997). Structural studies of jennite and 1.4 nm tobermorite: Disordered layering along the [100] of jennite. *Journal of the American Ceramic Society, 80*(12), 3021–3028.
57. Taylor, H. F. (1986). Proposed structure for calcium silicate hydrate gel. *Journal of the American Ceramic Society, 69*(6), 464–467.
58. Richardson, I. G. (2004). Tobermorite/jennite- and tobermorite/calcium hydroxide-based models for the structure of C–S–H: Applicability to hardened pastes of tricalcium silicate, h-dicalcium silicate, Portland cement, and blends of Portland cement with blast-furnace slag, metakaol. *Cement and Concrete Research, 34*(9), 1733–1777.
59. Bonnaud, P. A., Ji, Q., Coasne, B., Pellenq, R. J. M., & Van Vliet, K. J. (2012). Thermodynamics of water confined porous calcium silicate hydrate. *Langmuir, 28*(31), 11422 11432.
60. Manzano, H., Moeini, S., Marinelli, F., van Duin, A. C. T., Ulm, F. J., & Pellenq, R. J. M. (2011). Confined water dissociation in microporous defective silicates: Mechanism, dipole distribution, and impact on substrate properties. *Journal of the American Chemistry Society, 134*(4), 2208–2215.
61. Manzano, H., Masoero, E., Arbeloa, I. L., & Jennings, H. M. (2013). Molecular modelling of shear deformations in ordered and disordered calcium silicate hydrates. *Soft Matter, 9*(30), 7333–7341.

Chapter 3
Introduction to Simulation Techniques on the Cement-Based Materials

3.1 Introduction to the Molecular Simulation Method

Molecular simulation includes both theoretical methods and computational techniques to model the various properties of the molecules. The techniques have been widely applied in the fields of computational chemistry, drug design, computational biology, and materials science. With the development of science and technology, the modern computer provided strong support on the molecular simulation. Interaction of liquids by Monte Carlo (MC) was first studied by the assistance of computer in 1953 [1] and 3 years later, the molecular dynamics (MD) simulation was extended to hard spheres system [2]. The technique gained popularity in materials science since the 1970s. Based on experimental constraints from X-ray crystallography or NMR spectroscopy, molecular simulation was widely applied to refine molecular structure of the protein and crystals. For example, biological folding process was first simulated by MD in 1975, which paved the way for modern computational protein-folding [3]. In the field of physics, the molecular simulation helps explain the phenomena that cannot directly be obtained from experiments.

Molecular modeling plays a role in bridging the microscopic length and time scales and the macroscopic world of the laboratory. The simulation acts as a bridge in another sense between theory and experiment. A theory is tested by conducting a simulation using the same model, which can be used to compare with the experimental results. There are various techniques utilized in molecular modeling. In our current study, molecular mechanics (MM), MD, and grand canonical Monte Carlo (GCMC) have been employed in simulating the atomistic behavior of cement-based materials at nanolevel.

© Science Press and Springer Nature Singapore Pte Ltd. 2020
D. Hou, *Molecular Simulation on Cement-Based Materials*,
https://doi.org/10.1007/978-981-13-8711-1_3

3.2 Molecular Mechanics

This method uses traditional classical mechanics to model molecular systems. Classical mechanics is used to describe the motion of macroscopic objects according to Newton's second law. There are two basic assumptions for the molecular mechanics simulations:

1. The atom is treated as the particle that is assigned a radius (typically the van der Waals radius), polarizability, and a constant net charge (generally derived from quantum calculations and/or experiment).
2. Bonded interactions are treated as "springs" with an equilibrium distance equal to the experimental or calculated bond length. The interaction between atoms can be calculated as a function of atomic position only with the electrons around them in an optimal distribution through effective simple analytical functions called "potential functions" or "force fields" [4].

The system energy in Molecular Mechanics is the sum of all interatomic interactions. The explicit functional forms between two atoms or a group of atoms are usually chosen based on physical insights to the nature of covalent or non-covalent bonding between atoms. The exact parameters of the functions are fitted to ab initio or experimental results. The most common functional forms that are applicable in cement and cement-related systems are briefly reviewed in the following section.

3.2.1 Potential Forms

3.2.1.1 ClayFF Force Field

ClayFF is a force field suitable for molecular simulation of hydrated crystalline compounds and their interfaces with liquid phases [5]. It is based on an ionic-covalent description of metal–oxygen interactions associated with hydrated phases. The ClayFF force field has already been used to successfully model the structures of oxide and hydroxide materials, the interactions of aqueous species with oxide and hydroxide surfaces, and the behavior of water and ionic species in the interlayers of layer-structure phases [6, 7]. Previous studies on different cement phases, including portlandite, AFm (hydrated calcium aluminates) and tobermorite, have confirmed that the ClayFF force field is suitable for calcium silicate system simulation, even though it is not initially parameterized for a cement system.

To represent water, ClayFF considers the flexible simple point charge (SPC) water model [8]. The covalent bonds between O–H can be represented by a functional form known as the Harmonic Potential:

$$U_{ij}{}^{HB} = \frac{1}{2}K_b(r - r_0)^2 \tag{3.1}$$

where K_b is the stretching force constant. Since the bonding functional form describes in an effective way all the different contributions to the chemical bond, usually the Coulombic interactions are subtracted within a bonded group. Similar to harmonic bonding terms, a three-body potential representing the H–O–H angle energy is usually defined by

$$U_{ij}{}^{HA} = \frac{1}{2} K_a (\theta - \theta_0)^2 \tag{3.2}$$

where θ_0 is the optimum angle between the three atoms and K_a is the bending force constant.

In ClayFF, all atoms are represented as point charges and are allowed to complete translational freedom. Metal–oxygen interactions are based on a simple 12-6 Lennard-Jones potential combined with Coulombic interactions. The total potential interaction energy of the simulated system is calculated as a sum of an electrostatic term for all Coulomb interactions between partial atomic charges and a Lennard-Jones (12-6) term modeling the short-range van der Waals dispersive interactions:

$$U = \sum_{i,j} \left\{ \frac{q_i q_j}{4\pi \varepsilon_0 r} + 4\varepsilon_{ij} \left[\left(\frac{\sigma_{ij}}{r_{ij}} \right)^{12} + \left(\frac{\sigma_{ij}}{r_{ij}} \right)^{6} \right] \right\} \tag{3.3}$$

where r_{ij} is the distance between atoms i and j, q_i and q_j are partial charges centered on these atoms, σ and ε are parameters of the Lennard-Jones interaction potential, and ε_0 is the dielectric permittivity of vacuum (equal to 8.85419×10^{-12} F/m). The Lennard-Jones parameters combination of different ions interaction can be obtained by the arithmetic rules:

$$\sigma = \frac{\sigma_{ii} + \sigma_{jj}}{2} \tag{3.4}$$

$$\varepsilon_{ij} = \sqrt{\varepsilon_{ii}\varepsilon_{jj}} \tag{3.5}$$

3.2.1.2 CSHFF Force Field

Even though the molecular structure of the C–S–H analogue can be accurately predicted, the elastic properties of the mineral computed from the ClayFF force field are underestimated in compared with the ab initio calculation. In order to overcome these deficiencies of the simple point charge potential, a new force field potential, CSHFF force field [9] was re-parameterized to calibrate both structural data and elasticity data. The fitting parameter process by considering the ClayFF set of short-range parameters and let all charges adjust (except those of water species) by using the potential fitting procedure within the GULP code. It should be noted that a distinction between interlayer calcium (Ca_w) and intra-layer ones (Ca_s) based on quantum-derived charges. By fitting data of the GGA-DFT cell parameters, the bulk and shear moduli and the entire GGA-DFT elastic tensor of 11 Å tobermorite, 29 parameters including the partial charge of atoms and Lennard Jones interactions are obtained. Even though the potential forms of the CSHFF and ClayFF are similar, the long-term coulombic interaction and short-range Lennard Jones interaction have been changed significantly. The force field has been widely used in C–S–H simulations and proved to be able to describe the structure, energy, and mechanical properties of various calcium silicate phases well [10–14].

3.2.1.3　ReaxFF Force Field

ReaxFF plays a role in bridging the gap between quantum chemical (QC) and empirical force field (EFF)-based computational chemical methods (Fig. 3.1). Even though QC methods are in general applicable to all chemical systems, regardless of connectivity, the computational expense makes them inapplicable for large systems with atoms more than 100. EFF methods describe the relationship between energy and geometry with a set of relatively simple potential functions. As shown in the ClayFF and CSHFF, EFF methods describe molecular or condensed phase systems by simple harmonic equations that describe the stretching and compression of bonds and the bending of bond angles, usually augmented by van der Waals potential functions and Coulomb interactions to describe nonbonded interactions. Due to relative computational simplicity, the EFF force field is efficient in simulating the large system with millions of particles. EFF methods have been very successful in describing physical interactions in and between molecules and condensed phase systems. The force field obtained from the fitting procedure reduces its reliability as compared with the original QC description, which restricts the transferability of the EFF method. It is the reason that these EFF methods cannot describe reactive systems, and in most cases the shape of the potential functions applied in these methods, like the aforementioned harmonic description of the bond length/bond energy relationship, would make it impossible to find parameter values that accurately describe bond energy toward the dissociation limit.

The concept of bond order/bond energy relation, as first formulated by Tersoff [16], allows for the construction of EFF methods that can, in principle, handle connectivity changes. This concept was used by Brenner [17], to construct the REBO-potential, an EFF method for hydrocarbon systems, allowing, for the first time, dynamical simulations of reactions in large ($\gg$100 atoms) systems. Over the years, REBO has enjoyed widespread application, but its transferability is limited as it is based on a relatively small training set and because of its exclusion of all non-bonded interactions.

As with the Brenner potential, a bond order/bond energy relationship lies at the center of the ReaxFF-potential. Bond orders are obtained from interatomic distances (Fig. 3.2) and are continually updated at every MD or energy minimization (MM) iteration, thus allowing for connectivity changes. These bond orders are incorporated in all valence terms (i.e., energy contributions dependent on connectivity, like valence

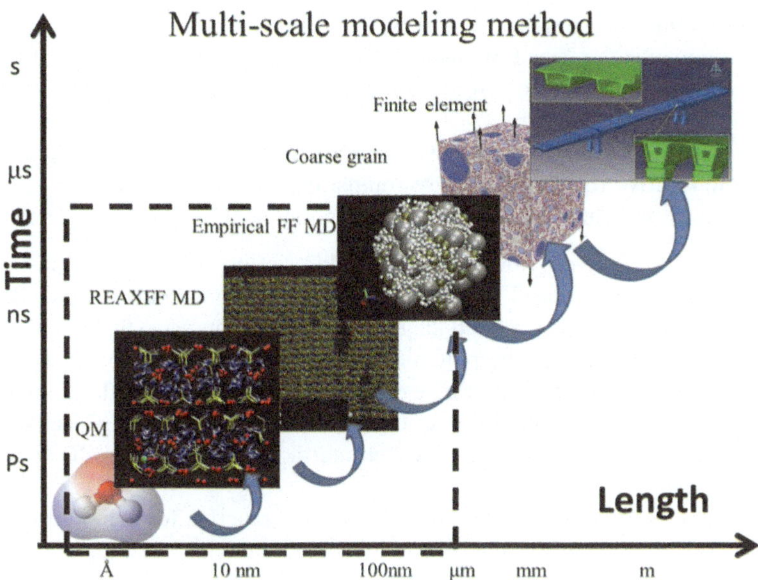

Fig. 3.1 Position of ReaxFF in the computational chemical hierarchy [15]

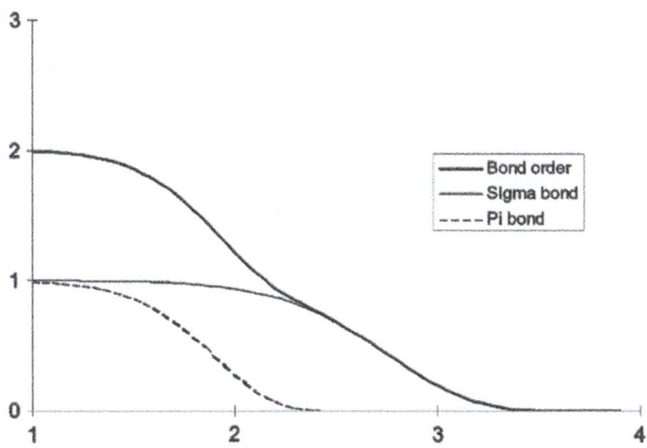

Fig. 3.2 Dependence of bond order on the interatomic distance [15]

angle and torsion angle energy) ensuring that energies and forces associated with these terms go to zero upon dissociation. Furthermore, ReaxFF describes non-bonded interactions between all atoms, irrespective of connectivity. Excessive short-range repulsive/attractive non-bonded interactions are circumvented by the inclusion of a shielding term in the van der Waals and Coulomb interaction.

ReaxFF aims to provide a transferable potential, applicable to a wide range of chemical environments. To ensure its transferability, the following general guidelines have been adopted in its development:

1. No discontinuities in energy or forces, even during reactions.
2. Each element is described by just one force field atom type. The ReaxFF metal oxide oxygen is described by the same parameters as the ReaxFF oxygen in organic molecules.
3. No pre-definition of reactive sites is necessary using ReaxFF. Although it is possible to drive reactions using restraints, this is not required; given the right temperature and chemical environment reactions will happen automatically.

The Reactive force field provides an advanced description of the interaction between Ca, Si, O, and H atoms in the C–S–H gel. The short-range interactions for the ReaxFF force field are determined by a bond length/bond order scheme so that the bonds can be broken and formed, with the potential energy transforming into a smooth state [17]. On the other hand, the long-range coulombic interactions are determined by a 7th order taper function, with an outer cut off radius of 10 Å. The Reactive Force field has been widely utilized in silica–water interfaces [18], calcium silicate hydrate gel [19] and nanocrystals [20].

Similar to empirical nonreactive force fields, the reactive force field divides the system energy up into various partial energy components.

$$E_{system} = E_{bond} + E_{over} + E_{under} + E_{lp} + E_{val} + E_{pen} + E_{tors}$$
$$+ E_{conj} + E_{vdwaals} + E_{coulomb} \tag{3.6}$$

The fundamental difference between ReaxFF and empirical force fields is that ReaxFF does not use fixed connectivity assignments for the chemical bonds. Instead the bond order, BO', is calculated directly from the instantaneous interatomic distances r_{ij} in Eq. (3.7), which are updated continuously. This allows for the creation and dissociation of bonds during a simulation. The bond energy (E_{bond}) is determined solely from BO as in Eq. (3.8). As the bond order BO turns to zero, the bond energy gradually transforms to zero, implying no discontinuity of the bond dissociation.

$$BO'_{ij} = BO'^{\sigma}{}_{ij} + BO'^{\pi}{}_{ij} + BO'^{\pi\pi}{}_{ij} = \exp\left[p_{bo,1}\left(\frac{r_{ij}}{r_0^{\sigma}}\right)^{p_{bo,2}}\right]$$
$$+ \exp\left[p_{bo,3}\left(\frac{r_{ij}}{r_0^{\pi}}\right)^{p_{bo,4}}\right] + \exp\left[p_{bo,5}\left(\frac{r_{ij}}{r_0^{\pi\pi}}\right)^{p_{bo,6}}\right] \tag{3.7}$$

$$E_{bond} = -D_e^{\sigma} BO_{ij}^{\sigma} \exp[p_{be,1}(1 - (BO_{ij}^{\sigma})^{p_{be,2}})] - D_e^{\pi} BO_{ij}^{\pi} - D_e^{\pi\pi} BO_{ij}^{\pi\pi} \tag{3.8}$$

Lone pairs on heteroatoms such as oxygen atoms can affect dramatically the response of these atoms to over- and under-coordination. Furthermore, the presence of these lone electron pairs influences the valence angles around atoms. In addition, by delocalizing, they can contribute to the stability of conjugated systems. Equation (3.9) describes the deviation of the number of lone pairs around an atom from the number of lone pairs at normal coordination (2 for oxygen, 0 for silicon and hydrogen).

$$E_{lp} = \frac{p_{lp}\Delta_{lp,1}}{1 + \exp(-75\Delta_{lp,1})} \tag{3.9}$$

For an over-coordinated atom ($\Delta > 0$), (3.10) imposes an energy penalty on the system. The form of (3.10) ensures that E_{over} will quickly vanish to zero for under-coordinated systems ($\Delta < 0$).

$$E_{over} = p_{lp} \cdot \Delta_l \cdot \frac{1}{1 + \exp(\Delta_l \cdot \lambda_6)} \tag{3.10}$$

For an under-coordinated atom ($\Delta < 0$), we want to take into account the energy contribution for the resonance of the π-electron between attached under-coordinated atomic centers. This is described in Eq. (3.11):

$$E_{under} = -p_{under} \cdot \frac{1 - \exp(\lambda_7 \cdot \Delta_i)}{1 + \exp(-\lambda_8 \cdot \Delta_j)} \cdot f_6(BO_{ij,\pi} \cdot \Delta_j) \tag{3.11}$$

Just as for bond terms, it is important that the energy contribution from valence angle terms goes to zero as the bond orders in the valence angle goes to zero. Equation (3.12) is used to calculate the valence angle energy contribution.

$$E_{val} = f_7(BO_{ij}) \cdot f_7(BO_{jk}) \cdot f_8(\Delta_j) \cdot \{k_a - k_a \exp[-k_b(\Theta_o - \Theta_{ijk})^2]\} \tag{3.12}$$

To reproduce the stability of systems with two double bonds sharing an atom in a valency angle, like allene, an additional energy penalty, as described in Eq. (3.13), is imposed.

$$E_{pen} = \lambda_{19} \cdot f_9(\Delta_j) \cdot \exp[-\lambda_{20} \cdot (BO_{ij} - 2)^2] \cdot \exp[-\lambda_{20} \cdot (BO_{jk} - 2)^2] \tag{3.13}$$

Just as with angle terms we need to ensure that dependence of the energy of torsion angle ω_{ijkl} accounts properly for BO smaller than 0 and for BO greater than 1. This is done by Eq. (3.14).

$$\begin{aligned}
E_{tors} = & f_{10}(BO_{ij}, BO_{jk}, BO_{kl}) \cdot \sin\Theta_{ijk} \cdot \sin\Theta_{jkl} \\
& \times \left[\frac{1}{2} V_2 \cdot \exp\{p_l(BO_{jk} - 3 + f_{11}(\Delta_j, \Delta_k))^2\} \right. \\
& \left. \cdot (1 - \cos 2\omega_{ijkl}) + \frac{1}{2} V_3 \cdot (1 + \cos 3\omega_{ijkl}) \right]
\end{aligned} \tag{3.14}$$

Equation (3.15) describes the contribution of conjugation effects to the molecular energy

$$E_{conj} = f_{12}(BO_{ij}, BO_{jk}, BO_{kl}) \cdot \lambda_{26} \cdot \left[1 + (\cos^2\omega_{ijkl} - 1) \cdot \sin\Theta_{ijk} \cdot \sin\Theta_{jkl}\right] \tag{3.15}$$

In addition to valence interactions which depend on overlap, there are repulsive interactions at short interatomic distances due to Pauli principle orthogonalization and attraction energies at long distances due to dispersion. These interactions, comprised of van der Waals and Coulomb forces, are included for all atom pairs, thus avoiding awkward alterations in the energy description during bond dissociation. In this respect, ReaxFF is similar in spirit to the central valence force fields that were used earlier in vibrational spectroscopy. To account for the van der Waals interactions we use a distance-corrected Morse-potential (Eq. 3.16). By including a shielded interaction (Eq. 3.16), excessively high repulsions between bonded atoms (1–2 interactions) and atoms sharing a valence angle (1–3 interactions) are avoided.

$$E_{vdWaals} = D_{ij} \cdot \left\{ \exp\left[\alpha_{ij} \cdot \left(1 - \frac{f_{13}(r_{ij})}{r_{vdw}} \right) \right] - 2 \cdot \exp\left[\frac{1}{2} \cdot \alpha_{ij} \cdot \left(1 - \frac{f_{13}(r_{ij})}{r_{vdw}} \right) \right] \right\}$$

$$(3.16)$$

As with the van der Waals interactions, Coulomb interactions are taken into account between all atom pairs. To adjust for orbital overlap between atoms at close distances a shielded Coulomb potential is used.

$$E_{coulomb} = C \cdot \frac{q_i \cdot q_j}{\left[r_{ij}^3 + \left(\frac{1}{\gamma_{ij}} \right)^3 \right]^{1/3}} \tag{3.17}$$

Atomic charges are calculated using the electron equilibration method (EEM) approach [21].

3.2.2 Energy Minimization

Energy minimization is to find the most stable configuration of molecules. The energy minimization algorithms, Steepest Descent and Conjugate Gradients method [22], are introduced. These algorithms are derivative-based, meaning that these techniques employ the derivatives of the potential energy. The negative first derivative of the potential energy with respect to interatomic distance (r) is the force (F) acting on an atom. The potential energy of a system may be expressed by a Taylor series expansion as follows:

$$F = -\frac{\partial U}{\partial r} \tag{3.18}$$

$$U(r + \delta r) = U(r) + \frac{\partial U}{\partial r} \delta r + \frac{1}{2!} \frac{\partial^2 U}{\partial r^2} (\delta r)^2 + \cdots \tag{3.19}$$

The potential energy expression is usually truncated to either first derivative, also called the gradient vector (g) or second derivatives, termed as hessian matrix (H).

In steepest descent methods, Eq. (3.19) is approximated to the first derivative of the potential energy. For every iterative step of this algorithm a line search or an arbitrary step size is used to determine the direction of steepest decent. The net force acting on individual atoms is computed from the potential energy expression and the atomic positions are computed according to Eq. (3.20).

$$r_i^j = r_i^{j-1} + \alpha \cdot F_i \tag{3.20}$$

The procedure is repeated for every time step (j) until the force (F) on the individual atoms reaches zero. Where i is the atom number, j is the time step, α is the multiplication factor, and F_i is the net force on an atom at that time step. In this method, the successive step directions in the iterative process are perpendicular to each other (as shown in Fig. 3.3).

The steepest descent method is quite effective when the initial configuration is far away from the minimum energy configuration, i.e., when the gradient of the potential energy is large. This method will not be efficient when it is close to approaching a minimum configuration because of the smaller gradient of the potential energy. However, these algorithms are known to have great numerical stability for the fact that

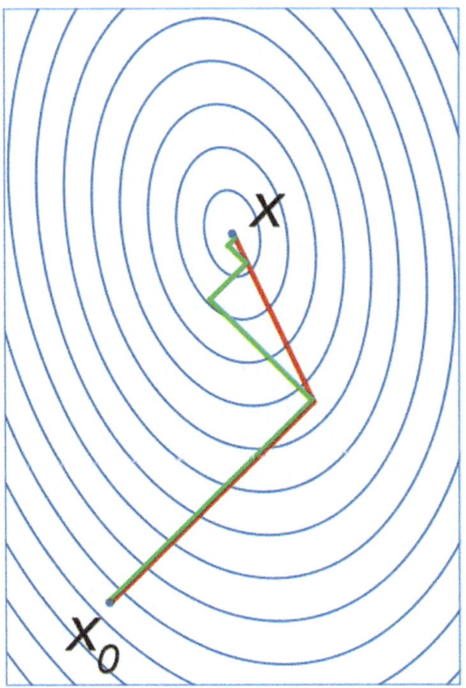

Fig. 3.3 A comparison of the convergence of gradient descent with optimal step size (in green) and conjugate vector (in red) for minimizing a quadratic function associated with a given linear system

the potential energy of the system could never increase for a reasonable assumption of α. Hence, this could be used along with the conjugate gradient method, a relatively efficient and quicker method.

The conjugate gradient method is an effective method when the gradient of the potential energy is smaller. In this method, the succeeding search directions are made conjugate, i.e., the step directions are made orthogonal to its preceding search vector. This algorithm introduces a vector perpendicular to the direction of search and moves it in another direction which is orthogonal to that vector. The residuals calculated in the method of conjugate gradient methods (as shown in Fig. 3.3) are orthogonal to preceding directional search vectors. This property warrants production of fresh and independent directional search vector throughout the iteration process until the residue becomes zero. This method finds the minima in lesser number of steps compared to the steepest gradient methods.

3.2.3 Elastic Properties

The calculation of elastic constants is very useful considering the fact that experimental determinations of these values are practically difficult. The elastic constants (C_{ij}) are determined by calculating the second derivatives of the energy density (energy/volume) with respect to lattice strain components.

$$C_{ij} = \frac{1}{V}\left(\frac{\partial^2 U}{\partial \varepsilon_i \varepsilon_j}\right) \tag{3.21}$$

The six possible strains denoted by a 6×6 symmetric matrix represents the hardness of the material with respect to the deformations [23]:

$$\mathbf{C}_{ij} = \begin{bmatrix} C_{11} & C_{12} & C_{13} & C_{14} & C_{15} & C_{16} \\ C_{21} & C_{22} & C_{23} & C_{24} & C_{25} & C_{26} \\ C_{31} & C_{32} & C_{33} & C_{34} & C_{35} & C_{36} \\ C_{41} & C_{42} & C_{43} & C_{44} & C_{45} & C_{46} \\ C_{51} & C_{52} & C_{53} & C_{54} & C_{55} & C_{56} \\ C_{61} & C_{62} & C_{63} & C_{64} & C_{65} & C_{64} \end{bmatrix} \tag{3.22}$$

The elastic compliance matrix, S, is obtained by finding the inverse matrix of C (i.e., $S = C^{-1}$).

The elastic constants and compliance can be further utilized to compute the bulk modulus, shear modulus and Young's modulus in Voight–Reuss–Hill expressions [24]:

$$K_{Voight} = \frac{1}{9}(C_{11} + C_{22} + C_{33} + 2(C_{12} + C_{13} + C_{23})) \tag{3.23}$$

$$K_{Reuss} = (S_{11} + S_{22} + S_{33} + 2(S_{12} + S_{13} + S_{23}))^{-1} \tag{3.24}$$

$$G_{Voight} = \frac{1}{15}(C_{11} + C_{22} + C_{33} + 3(C_{44} + C_{55} + C_{66}) - C_{12} - C_{13} - C_{23}) \tag{3.25}$$

$$G_{Reuss} = \frac{15}{4(S_{11} + S_{22} + S_{33} - S_{12} - S_{13} - S_{23}) + 3(S_{44} + S_{55} + S_{66})} \tag{3.26}$$

The Hill modulus is averaged between the Voight and Reuss one. Based on the bulk modulus and shear modulus, the Poisson ratio and Young's modulus can be described in the following equations:

$$v = \frac{3K - 2G}{6K + 2G} \tag{3.27}$$

$$E = 2G(1 + v) \tag{3.28}$$

3.3 Molecular Dynamics

MD is a numerical computational tool that describes the physical movements of atoms in the given system through a period of time [25]. Figure 3.4 gives a clear illustration of the basic principle and procedure of the MD simulation. MD gives the motion of the particles based on Newton's 2nd law at finite temperatures and the interaction between atoms and the potential energy is defined by the molecular mechanics mentioned in Sect. 3.2. For systems which obey the ergodic hypothesis [26], the evolution of a single molecular dynamics simulation may be used to determine macroscopic thermodynamic properties of the system: the time averages of an ergodic system correspond to micro-canonical ensemble averages. According to the statistical mechanics, averaging the trajectories, positions, and velocities of all atoms generated by the MD at microscale can obtain the macroscopic properties. In this respect, MD simulation acts as a bridge between microscopic length and time scales and macroscopic world of laboratory: it provides a hypothesis of the interactions between molecules and predicts the "exact" bulk properties.

3.3.1 Ensembles

The microscopic state of a system is defined by the atomic positions and momenta; which are the coordinates in a multidimensional imaginary space called phase space. For a given system, any state can be entirely described by the positions, q, and momentum, p, of all the particles within the system.

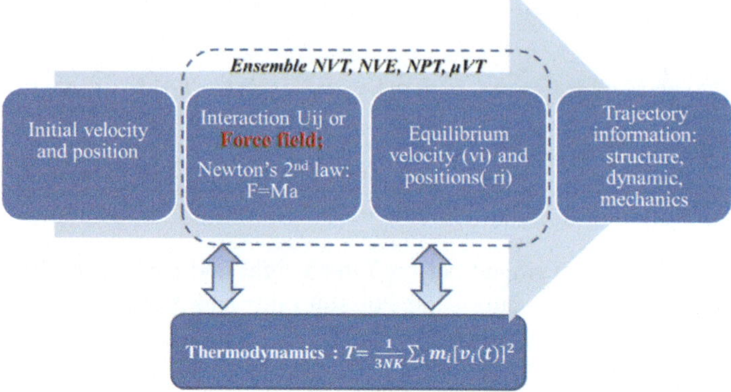

Fig. 3.4 The fluid chart of molecular dynamics simulation procedure and basic principle

$$q = (q_1 q_2 \ldots q_n) \text{ and } p = (p_1 p_2 \ldots p_n); \tag{3.29}$$

the two sets of vectors, q and p, represent each single point in the phase space whose evolution trajectory can be tracked. Most of the system thermodynamic properties in MD can be averaged over different ensembles using p and q. For instance, for a system property, say A, at equilibrium, the ensemble average reads

$$A = \iint^A (q, p) P(q, p) dp dq \tag{3.30}$$

In equation, $P(q, p)$ is the probability of the system to be at the phase space point (q, p), and $A(q, p)$ is the system property value at that point.

For a system of N particles, this space has 6N dimensions and hence a single point in phase space represents the state of a system. A molecular dynamic simulation generates a collection of points in phase space as a function of time. An ensemble can be defined as a collection of points in phase space satisfying the conditions of a particular thermodynamic state. There are three commonly used ensembles in MD:

Canonical ensembles: The volume, V, the number of molecules, N, and the temperature, T, are fixed in each system. The walls surrounding each system are impermeable to molecules but are perfectly thermally conducting, and the system is surrounded by a fictitious infinite heat reservoir at temperature T. The ensemble average of any quantity, F, that depends on the state of the system can be found from

$$< F > = \frac{\sum_i F_i \exp(-E_i(N, V)/k_b T)}{Q(N, V, T)}; \tag{3.31}$$

where $Q(N, V, T)$ is the canonical ensemble partition function:

$$Q(N, V, T) = \sum \exp\left(-\frac{E(N, V)}{k_b T}\right) \tag{3.32}$$

Micro-canonical ensemble: The macroscopic variables of the micro-canonical ensemble are quantities such as the total number of particles in the system (symbol: N), the system's volume (symbol: V) each which influence the nature of the system's internal states, as well as the total energy in the system (symbol: E). This ensemble is therefore sometimes called the NVE ensemble, as each of these three quantities is a constant of the ensemble.

Isothermal–isobaric (N, P, T): Number of atoms, Pressure, and Temperature of the system are kept constant.

3.3.2 MD Algorithm

The basic MD algorithm is demonstrated in Fig. 3.5. To begin a molecular dynamics simulation, an initial configuration of the system, a starting point, or $t = 0$ should be selected. The choice of the initial configuration must be done carefully as this can influence the quality of the simulation. It is often good to choose a configuration close to the state that you wish to simulate. As to the C–S–H analogue, the crystal structure of tobermorite 11 Å, obtained from the NMR and XRD experiment, is set as the initial structure.

Before starting a molecular dynamics simulation, it is necessary to operate energy minimization on the structure to remove any strong van der Waals or Coulombic interactions. Otherwise, it leads to local structural distortion and results in an unstable simulation. Subsequently, initial velocities at a low temperature are assigned to each atom of the system and Newton's equations of motion are integrated to propagate the system in time. To numerically solve the system of partial differential equations, many algorithms have been proposed. Perhaps the most common algorithm is due to Verlet [27] who proposed a simple, yet efficient method to integrate equation. In Verlet algorithm, the initial velocities are usually adjusted to a desired kinetic energy to conserve the zero value of the momentum. Next, both the backward ($t - \Delta t$) and forward ($t + \Delta t$) Taylor expansion of the particles coordinates are written around the time step, Δt, up to the fourth term. Summing both backward and forward Taylor expansions gives

$$q(t + \Delta t) \approx 2q(t) - q(t - \Delta t) + \frac{F(t)}{m}\Delta t^2 \pm O(\Delta t^4) \tag{3.33}$$

where F, m, and t are the forces acting on the particle, mass of the particle, and the current time, respectively. In Verlet algorithm, to calculate the next positions, only the past and current positions are needed and not the velocities. However, analogous to positions, velocities can be obtained from the rest of the Taylor expansion for both backward and forward directions:

Fig. 3.5 Flow chart of the MD algorithm

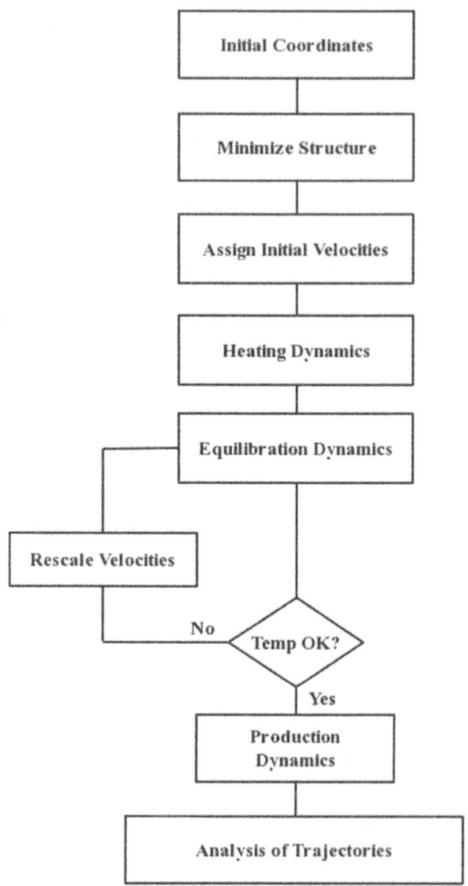

$$v(t) \approx \frac{q(t + \Delta t) - q(t - \Delta t)}{2\Delta t} \pm O(\Delta t^2) \tag{3.34}$$

During the heating process, initial velocities are assigned at a low temperature and the simulation is started. Due to thermodynamic constraint by the ensemble, new velocities are assigned at a slightly higher temperature and the simulation is allowed to continue. This is repeated until the desired temperature is reached. Once the desired temperature is reached, the simulation system continues and during this process several properties are monitored. The point of the equilibration phase is to run the simulation until these properties become stable with respect to time. If the temperature increases or decreases significantly, the velocities can be scaled such that the temperature returns to near its desired value. The final step of the simulation is to run the simulation in "production" phase for the time length desired. This can be from several hundred ps to ns or more. It is during the production phase that thermodynamic parameters can be calculated so the simulation must conform to one of the ensembles described earlier.

When carrying out an MD simulation, coordinates, velocities, and energy of the system are saved; these are then used for the analysis. Time-dependent properties can be displayed graphically, where one of the axis corresponds to time and the other to the quantity of interest, such as energy, MSD, and so on. The average potential energy U is obtained by averaging its instantaneous value, which is usually obtained straightforwardly at the same time as the force computation is made. Knowledge of average potential energy U is required to verify energy conservation. This is an important check to do in any MD simulation. The instantaneous kinetic energy is given by:

$$K(t) = \frac{1}{2} \sum_i m_i [v_i(t)]^2 \tag{3.35}$$

The temperature, related to the kinetic energy, can be calculated by:

$$K = \frac{3}{2} N k_B T \tag{3.36}$$

The pressure of the entire system of atoms is a function of the position and force on each atom. It is computed by the following equation:

$$P = \frac{N k_B T}{V} + \frac{\sum_i^N r_i \cdot f_i}{d V} \tag{3.37}$$

where N is the number of atoms in the system; K_B is the Boltzmann constant; T is temperature; d is the dimensionality of the system; and V is the system volume.

3.3.3 MD Trajectories Analysis

As mentioned in Fig. 3.5, MD simulation includes equilibrium run and production run. The trajectory analysis is based on the later procedure to establish the relation between the properties of the atoms and the trajectory information of atoms in the simulation system. The radial distribution function, mean square displacement, time correlation function, and uniaxial tension test are commonly utilized to characterize the molecular structure, dynamics, and mechanical behavior of the material.

Radial distribution function
The physical meaning of the radial distribution function can be shown in Fig. 3.6. Where the black sphere at the center is the reference molecule. Radial distribution function can be interpreted as the ratio of the local density to the bulk density of the system, which can be defined by the following equation:

Fig. 3.6 Schematic diagram
of radial distribution function

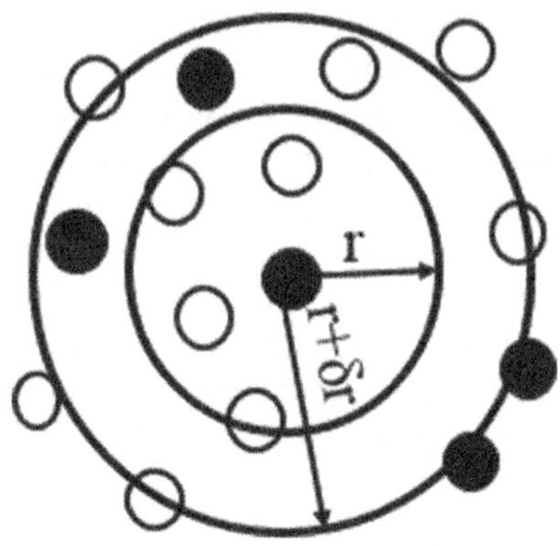

$$g(\mathrm{r}) = \frac{dN}{\rho 4\pi r^2 dr} \qquad (3.38)$$

where dN refers to the number of molecules in the range of dr distance, ρ refers to
the density of the system. Radial distribution functions show the distance distribution
of two different atoms or ions, and can give significant structural information and
illustrate the spatial correlation of atoms.

Mean square displacement

Mean square displacement MSD(t) [28], a parameter to estimate the dynamic properties of atoms, can be defined by the following equation:

$$\mathrm{MSD}(t) = <|r_i(t) - r_i(0)|^2> \qquad (3.39)$$

where $r_i(t)$ represents the position of atom i at time t, $r_i(0)$ is the original position
of atom i, and MSD takes into account threedimensional coordinates. A large MSD
value at time t indicates that the atoms diffuse rapidly and are displaced far away from
the original position. The self-diffusion coefficient ($\boldsymbol{D}$) is derived from the diffusive
regime from the MSD curves by the following equation:

$$2\mathrm{D} \cdot t = \frac{1}{3} <|\boldsymbol{r}_i(t) - \boldsymbol{r}_i(0)|^2> \qquad (3.40)$$

Time correlated function

The time correlated function (TCF) is utilized to describe pair dynamical properties
of atom–atom correlations, as well as the bonding stability between solution species
and atoms in the material. TCF of pairs in the simulation system is represented by
the following equation:

$$C(t) = \frac{<\delta b(t)\delta b(0)>}{<\delta b(0)\delta b(0)>} \qquad (3.41)$$

where $\delta b(t) = b(t) - \,<b>$, $b(t)$ is a binary operator that takes a value of one if an ion–water pair is within the nearest neighbor separation (in the hydration shell) at time t and zero otherwise, and $<b>$ is the average value of b overall simulation time and pairs. The border of the hydration shell is defined as the first minimum of ion–oxygen radial distribution function. For the species in the solution, the meaning of **TCF** is the probability that a water molecule or ion, which was in the hydration shell of ion i initially, is also in the hydration shell of the same at time t. TCF for the solution species and surface atoms indicates that a molecule or ion, which formed chemical bonds with surface atom i at the very beginning, remains binding with the same atom at time i. The evolution of $C(t)$ describes the dynamics of ion–water pair structural relaxation and its relaxation time τ can be obtained by integrating the $C(t)$ function:

$$\tau_{res} = \int_0^\infty C_\theta(t)dt \tag{3.42}$$

Uniaxial Tension Test

Uniaxial tension testing is employed to investigate the mechanical behavior and fracture process of the material. Super-cells are obtained by periodically extending the simulation model for the uniaxial tension test. It should be noted that using a large number of atoms in super-cells can give stable statistical simulation results, especially in regard to reliable failure modes. In order to explore the failure mechanism of the material, the stress–strain relation and the change of molecular structure are investigated in the loading process.

To obtain the stress–strain relation, the structure is subjected to uniaxial tensile loading through gradual elongation at constant strain rates usually in the range from 0.08/ps to 0.008/ps. In the whole simulation process, NPT ensembles are defined for the system. Taking the tension along x-direction at 300 K for example, the super-cells were firstly relaxed at 300 K and coupled to zero external pressure in the x, y, z dimensions for 500 ps. Then, after the pressures in the three directions reached equilibrium, the structure would be elongated in the x-direction. Meanwhile, the pressure in y, z-direction was kept at zero. Pressure evolution in the x-direction was taken as the internal stress σ_{xx}. Setting the pressure perpendicular to the tension direction to zero can allow the normal direction to relax un-isotropically without any restriction. The setting, considering Poisson's ratio, can eliminate the artificial constraint for the deformation.

The stress tensor component was calculated by the following equation:

$$P_{IJ} = \frac{\sum_k^N m_k v_{kI} v_{kJ}}{V} + \frac{\sum_k^N r_{kI} f_{kJ}}{V} \tag{3.43}$$

where V is the simulation box volume, I and J are equal to x, y, and z; m_k, v_{kI}, v_{kJ}, and f_{kJ} are the I components of the momentum, position, and force acting on the kth atom with mass m_k.

3.4   Grand Canonical Monte Carlo (GCMC)

The ensembles we have discussed so far have the total number of particles imposed. However, in some adsorption studies, one would like to know the amount of material adsorbed as a function of the pressure and temperature of the reservoir with which the material is in contact. Since the adsorption process simulated by the MD techniques is computational expensive, most cases of the adsorption process are investigated by the grand canonical Monte Carlo (GCMC) method or μVT ensemble, in which the temperature, volume, and chemical potential are fixed [25]. In the experimental setup, the adsorbed gas is in equilibrium with the gas in the reservoir. The equilibrium conditions are that the temperature and chemical potential of the gas inside and outside the adsorbent must be equal. The gas that is in contact with the adsorbent can be considered as a reservoir that imposes a temperature and chemical potential on the adsorbed gas (see Fig. 3.7). Therefore, only the temperature and chemical potential of this reservoir determine the equilibrium concentration inside the adsorbent.

To generate a Markov chain in the grand canonical ensemble, two new moves should be added that generate trial states with differing number of molecules. Thus, at every step in the chain, moving a molecule, creating a new molecule or destroying an existing molecule should be decided randomly. In order to satisfy the symmetry in the underlying transition matrix, it is required that the probability of creating a molecule is the same as that of destroying it immediately. The acceptance/rejection probability in Metropolis algorithm is shown in Eqs. (3.44) and (3.45) [25].

$$P_{acc} = \min\left[1, \frac{V}{\Lambda(N+1)} \exp\left(-\frac{(\mu - E_N(N+1) + E_M(N))}{k_B T}\right)\right] \quad (3.44)$$

where μ is the target chemical potential and N is the number of molecules while V is the system volume. Similarly, the acceptance condition for "deletion" of a molecule in μVT ensembles is

$$P_{acc} = \min\left[1, \frac{\Lambda^3 N}{V} \exp\left(-\frac{(\mu + U_{New}(N+1) - U_{old}(N))}{k_B T}\right)\right] \quad (3.45)$$

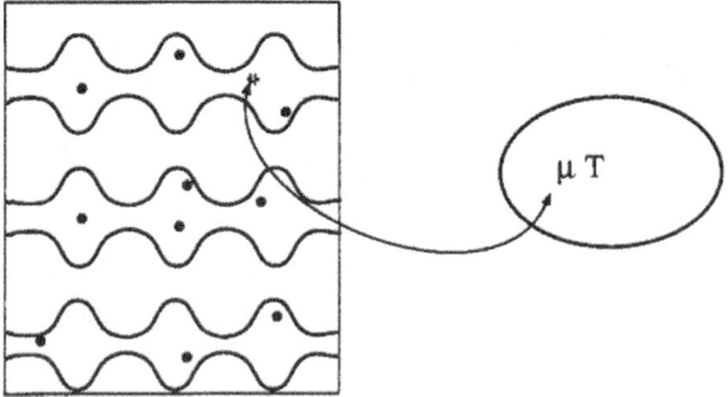

Fig. 3.7 Adsorbent in contact with a reservoir that imposes constant chemical potential and temperature by exchanging particles and energy

3.5 Chapter Summary

In this chapter, several molecular modeling techniques are introduced to simulate the molecular behavior of the cement-based material. Some conclusions can be drawn as follows:

(1) Foundational physical and chemical principles of molecular dynamics, Monte Carlo simulation and molecular mechanics have been briefly discussed. Generally, these methods play an essential role in constructing the cement model at nanoscale.

(2) Computational simulation acts as a bridge between microscopic length and time scales and macroscopic world of laboratory: it provides a hypothesis of the interactions between molecules and predicts the "exact" bulk properties.

(3) The empirical and reactive force fields are compared in the respect of transferability, computational efficiency, and applicable fields. The ReaxFF, the newly developed force field, provide some insights of chemical reaction, which cannot be defined by the CSHFF and ClayFF. The reactive force field mentioned above is taken as the major potential form not only in the cement model construction but in simulating the mechanical, dynamic, and thermodynamic properties of cement-based material.

(4) For studying the dynamical trajectories of a system at equilibrium state, MD provides useful information. On the other hand, MC methods, adjusting various ensemble, can be efficient to explore the properties of a system near equilibrium.

References

1. Metropolis, N., Rosenbluth, A. W., Rosenbluth, M. N., Teller, A. H., & Teller, E. (1953). Equation of state calculations by fast computing machines. *The Journal of Chemical Physics, 21*(6), 1087–1092.
2. Alder, B., & Wainwright, T. E. (1958). *Molecular dynamics by electronic computers* (pp. 97–131).
3. Levitt, M., & Warshel, A. (1975). Computer simulation of protein folding. *Nature, 253*(5494), 694.
4. Leach, B. A. R. (1996). *Molecular modelling: Principles and applications* (pp. 199–200). Longman.
5. Cygan, R. T., Liang, J.-J., & Kalinichev, A. G. (2004). Molecular models of hydroxide, oxyhydroxide, and clay phases and the development of a general force field. *Journal of Physical Chemistry B, 108*(4), 1255–1266.
6. Kirkpatrick, R. J., Kalinichev, A. G., Hou, X., & Struble, L. (2005). Experimental and molecular dynamics modeling studies of interlayer swelling: Water incorporation in kanemite and ASR gel. *Materials and Structures, 38*(4), 449–458.
7. Cygan, R. T., Greathouse, J. A., Heinz, H., & Kalinichev, A. G. (2009). Molecular models and simulations of layered materials. *Journal of Materials Chemistry, 19*(17), 2470–2481.
8. Berendsen, H. J. C., Grigera, J. R., & Straatsma, T. P. (1987). The missing term in effective pair potentials. *Journal of Physical Chemistry, 91*(24), 6269–6271.
9. Shahsavari, R., Pellenq, R. J., & Ulm, F. J. (2010). Empirical force fields for complex hydrated calcio-silicate layered materials. *Physical Chemistry Chemical Physics, 13*(3), 1002–1011.

10. Pellenq, J. M., Kushima, A., Shahsavari, R., Vliet, K. J. V., Buehler, M. J., Yip, S., et al. (2009). A realistic molecular model of cement hydrates. *PNAS, 106*(38), 16102–16107.
11. Youssef, M., Pellenq, R. J. M., & Yildiz, B. (2011). Glassy nature of water in an ultraconfining disordered material: The case of calcium–silicate–hydrate. *Journal of the American Chemical Society, 133*(8), 2499–2510.
12. Ji, Q., Pellenq, J. M., & Vliet, K. J. (2012). Comparison of computational water models for simulation of calcium–silicate–hydrate. *Computational Materials Science, 53*(1), 234–240.
13. Bonnaud, P. A., Ji, Q., Coasne, B., Pellenq, R. J., & Van Vliet, K. J. (2012). Thermodynamics of water confined in porous calcium-silicate-hydrates. *Langmuir, 28*(31), 11422.
14. Manzano, H., Moeini, S., Marinelli, F., van Duin, A. C., Ulm, F. J., & Pellenq, R. J. M. (2012). Confined water dissociation in microporous defective silicates: Mechanism, dipole distribution, and impact on substrate properties. *Journal of the American Chemical Society, 134*(4), 2208–2215.
15. van Duin, A. (2002). *ReaxFF user manual*. Materials and Process Simulation Center, California Institute of Technology.
16. Tersoff, J. (1986). New empirical model for the structural properties of silicon. *Physical Review Letters, 56*(6), 632.
17. Brenner, D. W., Shenderova, O. A., Harrison, J., Stuart, S. J., Ni, B., & Sinnott, S. B. (2002). A second-generation reactive empirical bond order (REBO) potential energy expression for hydrocarbons. *Journal of Physics: Condensed Matter, 14*(4), 783–802.
18. Leroch, S., & Wendland, M. (2012). Simulation of forces between humid amorphous silica surfaces: A comparison of empirical atomistic force fields. *Journal of Physical Chemistry C, 116*(50), 26247.
19. Manzano, H., Masoero, E., Lopezarbeloa, I., & Jennings, H. M. (2013). Shear deformations in calcium silicate hydrates. Soft Matter, 9(30), 7333–7341.
20. Lau, T. T., Kushima, A., & Yip, S. (2010). Atomistic simulation of creep in a nanocrystal. *Physical Review Letters, 104*(17), 175501.
21. Janssens, G. O. A., Baekelandt, B. G., Toufar, H., Mortier, W. J., & Schoonheydt, R. A. (1995). Comparison of cluster and infinite crystal calculations on zeolites with the electronegativity equalization method (EEM). *Journal of Physical Chemistry A, 99*(10), 3251–3258.
22. Knyazev, A. V., & Lashuk, I. (2007). Steepest descent and conjugate gradient methods with variable preconditioning. *SIAM Journal on Matrix Analysis and Applications, 29*(4), 1267–1280.
23. Wert, C. A., Thomson, R. M., & Armstrong, H. L. (1964). Physics of solids. *American Journal of Physics, 33*(5), 417.
24. Hill, R. (1952). The elastic behaviour of a crystalline aggregate. *Proceedings of the Physical Society A, 65*(5), 349–354.
25. Frenkel, D., Smit, B., & Ratner, M. A. (1996). *Understanding molecular simulation: From algorithms to applications* (p. 66). Academic Press, Inc.
26. Infeld, L. (1938). *The principles of statistical mechanics by Richard C. Tolman* (pp. 691–692). The Clarendon Press.
27. Verlet, L. (1967). Computer "experiments" on classical fluids. I. Thermodynamical properties of Lennard-Jones molecules. *Physical Review, 22*(1), 79–85.
28. Kerisit, S., & Liu, C. J. (2009). Molecular simulations of water and ion diffusion in nanosized mineral fractures. *Environmental Science & Technology, 43*(3), 777–782.

Chapter 4
Modeling the Calcium Silicate Hydrate by Molecular Simulation

Chapter 2 reviewed the experimental, theoretical, and computational study on the molecular structure of the C–S–H gel. Experimental study provides the physical and chemical features of the C–S–H gel, which provides fundamental base for the modeling. Meanwhile, the theoretical contributions give valuable insights into the molecular structural evolution mechanism of the complicated cement hydrate. Molecular simulation methods reviewed in Chap. 3, bridging the theoretical and experimental study, open a novel way for understanding the nanoscale structure of the cement hydrate. In this chapter, computational chemistry method is employed to construct the C–S–H model, and the newly constructed C–S–H model is both validated by the experimental testing and based on the basic principles of the crystal chemistry. Also, the validated model is utilized to predict the structure, dynamics, and mechanical properties of the cement hydrate.

4.1 Introduction

The molecular structure of C–S–H has been studied for more than half a century, particularly using a variety of experimental techniques, including NMR [1], XRD [2], and SANS [3], and C–S–H is now widely believed to be the analogue of layered minerals: tobermorite [4, 5] and jennite [6]. Based on experimental data, Pellenq et al. [7] employed the molecular dynamic (MD) method to construct a C–S–H model and is defined as a "realistic model". Meanwhile, a CSHFF force field [8] was developed to describe the interaction between atoms in cement systems, and this force field demonstrated good transferability in simulating cement-based materials. The model has also been widely applied to explain the glassy nature for the water molecules highly confined in the nanopores [9], the thermodynamic properties for the C–S–H gel [10], the comparison between different water models [11], and the alumina introduction in the C–S–H (CASH) [12].

However, there are lots of limitations for the "realistic model", which reduces the accuracy for the structure. To achieve the proper Ca/Si ratio, many Q_0 species,

© Science Press and Springer Nature Singapore Pte Ltd. 2020
D. Hou, *Molecular Simulation on Cement-Based Materials*,
https://doi.org/10.1007/978-981-13-8711-1_4

representing the monomer silicate structures, remained 13% in calcium silicate sheet, which is more than 3 times larger than the results obtained from the ^{29}Si and ^{17}O cross-polarized NMR test [1, 13]. The overestimation of the Q_0 species can influence the calcium silicate skeleton. Hence, in the "realistic model", the local structure of some calcium atoms, including the Ca–O bond distance and coordination number of the Ca atoms, is not reasonable in regard to the crystal chemistry of the calcium silicate phase [14]. Another structural limitation is that in the "realistic model", OH bonds are only in the water molecules and neither the Ca–OH bonds nor the Si–OH bonds are present in the interlayer region, which is not consistent with the detecting of the Ca–OH bonding by INS [15]. This limitation results from the fact that the CSHFF, the empirical force field, cannot allow "hydrolytic reaction" for water molecules confined in the C–S–H gel. Fortunately, the hydroxyl issue has been solved by using a reactive force field and DFT to dissociate the water molecules completely [16].

4.2 Computational Details

For the C–S–H gel, the model construction in the present study is based on the procedures that combine the method proposed by Pellenq et al. [7] and Manzano et al. [16]. Firstly, the layered analogue mineral of C–S–H, tobermorite 11 Å without water, was taken as the initial configuration for the C–S–H model [5, 17]. The molecular structure for the tobermorite is shown in Fig. 4.1a. Silicate chains were then broken to match the Q species distribution with $Q_1 = 73.9\%$, $Q_2 = 21.4\%$ and $Q_0 = 4.7\%$. The mean silicate chain length (MCL $= 2(Q_2/Q_1 + 1) = 2.58$) is consistent with the results obtained both from NMR testing [1] and molecular dynamics simulation [18]. It is worth noting that the Q_0 percentage is controlled to less than 5%, also matching well with experimental results [13]. Meanwhile, the Ca/Si ratios range from 0.7 to 2.3 with an average value of 1.7. In order to increase the Ca/Si ratio, some of the bridging SiO_2 units were first removed as proposed by Pellenq et al. [7]. More importantly, some dimmer structures (Si_2O_4) were also removed to satisfy the Q species distribution, especially for the low Q_0 percentage. After omission of the dimmer structure, to maintain charge neutrality, some oxygen atoms in the silicate chains have to remain in the form of dangling atoms. In this way, the dry calcium silicate skeleton can be obtained, as shown in Fig. 4.1b.

Subsequently, the grand canonical Monte Carlo (GCMC) method was utilized to investigate the structure of the dry calcium silicate skeleton emerged in water solution [10]. The dry sample obtained by tobermorite transformation was utilized for simulation. GCMC simulations determine the properties of the water molecules confined in the calcium silicate system at constant volume V in equilibrium with a fictitious infinite reservoir of liquid bulk water solution, imposing its chemical potential $\mu = 0$ eV and its temperature $T = 300$ K [7]. The simulation process is analogue to water adsorption in the microporous phases, such as calcium silicate hydrate and zeolite [19]. The simulation included 300,000 circles for the system to reach equilibrium followed by 100,000 circles for the production run. For each circle,

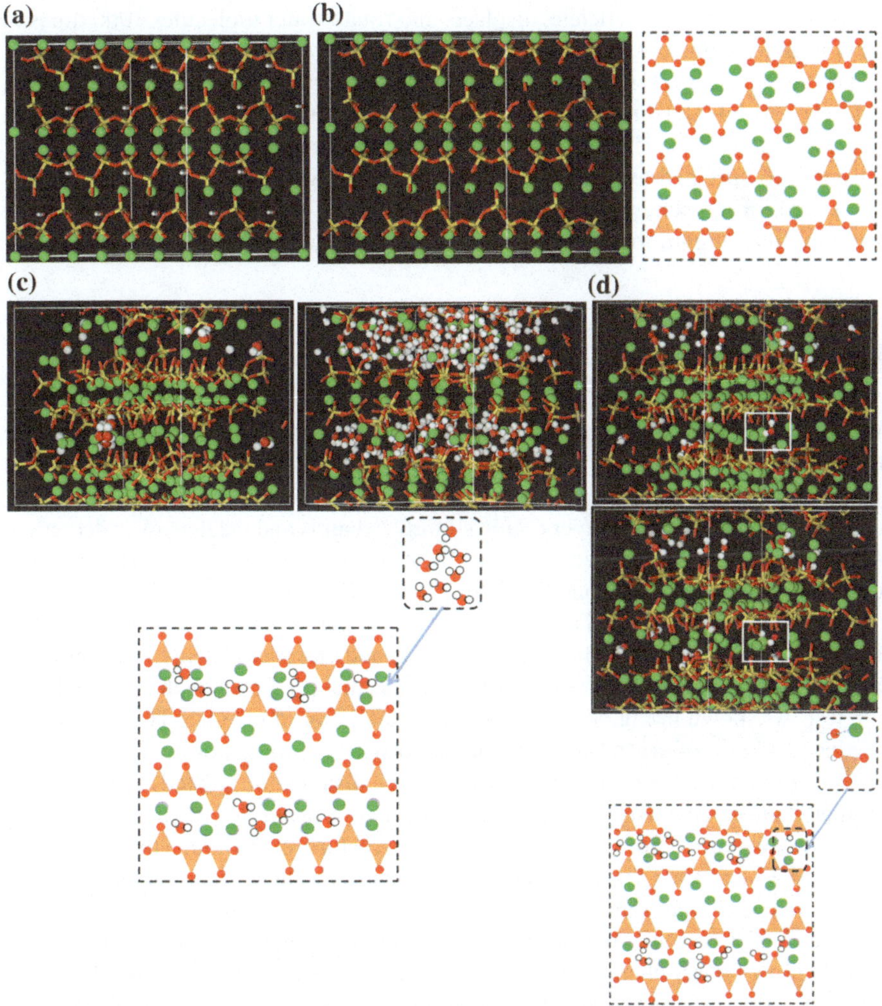

Fig. 4.1 **a** Super-cell $2 \times 3 \times 1$ of the crystal structure of tobermorite 11 Å. a = 22.32, b = 22.17, c = 22.77; $\alpha = 90°$, $\beta = 90°$, $\gamma = 90°$. **b** Initial dry C–S–H structure for water adsorption simulation. Simulation box size: a — 21.3 Å, b — 21.2 Å, c = 21.9 Å; $\alpha = 90°$, $\beta = 90°$, $\gamma = 90°$. **c** GCMC water adsorption process. **d** Adsorbed water molecules further dissociation. Yellow and red bond represents the silicate chain (Si–O); the green balls are corresponding to the calcium atoms and the white red sticks are water molecules

it was attempted to insert, delete, displace, and rotate water molecules 1000 times in the constant volume calcium silica hydrate system.

The final chemical formula of the saturated C–S–H structure in current simulation is $(CaO)_{1.69}(SiO_2) \cdot (H_2O)_{1.82}$, which is quite close to $(CaO)_{1.7}(SiO_2) \cdot 1.8H_2O$ obtained by the SANS test [3]. The reactive force field molecular dynamic simulations under constant pressure and temperature (NPT) for 300 ps give the structures of C–S–H gel at equilibrium states. A further 1000 ps NPT run was employed to achieve the equilibrium configuration for structural and dynamic analysis.

4.3 Experimental Validation of the C–S–H Model

At 0 K energy minimization, after the molecular structure is fully relaxed and the density is around 2.54 g/cm^3, which is close to the density obtained by the neutron scattering (2.6 g/cm^3). When the molecular model runs at 300 K by the molecular dynamics method, the interlayer region slightly expands and the density is decreased to 2.45 g/cm^3. The density reduction can be attributed to different behavior of structural water molecules at 0 and 300 K environment.

The structure of the model is validated by the properties calculated from the experiments, including the X-ray diffraction intensity and vibrational density measured by the infrared spectroscopy and the elastic modulus that predicted by the nanoindentation test. As shown in Fig. 4.2, the X-ray diffractogram of the C–S–H model shows a reduced degree of crystallinity as compared with the tobermorite. There exists a peak at around $7°$, which is a typical interlayer distance in layered minerals. The simulated curve is consistent with XRD information from the synthesized C–S–H structure. Regarding the respects of molecular structure, even though the local structure of Si–O and Ca–O demonstrate amorphous state, the C–S–H remains layered arrangement in longer range.

Spectra of the atomic motions in the transitional, librational, and vibrational frequency allow further characterization of the simulation model in dynamical respect. The power spectra are obtained from Fourier transformation of the atomic velocity autocorrelation functions. The experimental bands obtained from the infrared spectra and simulated intensities are plotted in Fig. 4.3 for comparison. Frequency ranges from 400 to 4000 cm^{-1}, in which the mineral demonstrates the unique feature. The band in the domain 440–450 cm^{-1} represents the deformation silicate tetrahedron. The band ranging from 660 to 670 cm^{-1} is contributed by the Si–O–Si bending and 810 and 970 cm^{-1} is attributed to the Si–O stretching in the silicate tetrahedron. In addition, infrared spectrum also provides useful information about the feature of water molecules. The band at about 1600 cm^{-1} is characterized by the H–O–H bending, while at 3300 cm^{-1} corresponds to the O–H stretching. However, the location of the simulated band slightly shifts to a larger distance. The discrepancy can be attributed to the glassy nature of water molecules confined by the calcium silicate sheets. Besides, water molecules decomposes to form Si–OH and Ca–OH, which

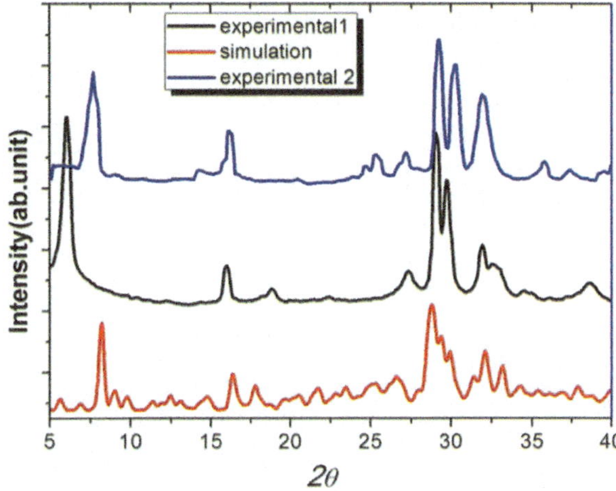

Fig. 4.2 X-ray data for simulated tobermorite (11 Å) and our C–S–H model. Experiment 1 is from the synthesized C–S–H in Ref. [20] and experiment 2 is about the tobermorite structure in [21]

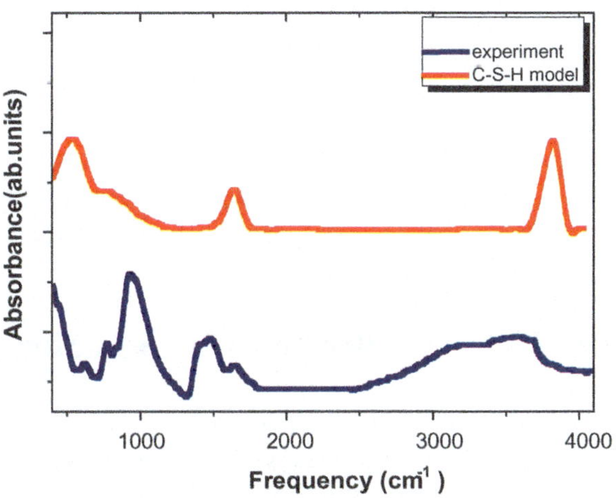

Fig. 4.3 Infrared data for simulated C–S–H model and from experimental result [22]

makes the bands of OH stretching and H–O–H bending more complicated. The feature of water and H-bonds connectivity will also discuss in the following section.

The elastic constants (C_{ij}) for the C–S–H model were calculated by using the box deformation method [11]. The imposed strain was 1%, and the strained structures were relaxed via conjugate gradient minimization prior to calculation of elastic constants. C_{ij} values, derived from the virial stress–strain relations of the deformed crystal cell, are listed in Table 4.1. Young's modulus in the present simulation is consistent with the results achieved from nanoindentation testing (60 GPa) [23] and previous ab initio calculations (55–68 GPa) proposed by Pellenq et al. [7]. The hardness of the C–S–H gel tested from the nanoindentation can reach as large as 4 GPa, which is consistent with the average tensile strength between the calcium silicate sheet and the weak interlayer connection (Table 4.2).

Table 4.1 Elastic tensor

C_{ij} (GPa)	1	2	3	4	5	6
1	99.15	34.7	24.02	−0.18	−0.66	0.11
2		109.8	33.2	1.19	−0.42	1.7
3			87.5	−0.55	−0.33	−1.15
4				22.9	0.046	−0.23
5					17.63	0.93
6						24.35

Table 4.2 Mechanical properties

Elastic properties	
Voigt bulk modulus	53.4 GPa
Reuss bulk modulus	52.4 GPa
Voigt shear modulus	26.6 GPa
Reuss shear modulus	24.8 GPa
Poisson's ratio	0.29
Young's modulus	66.4 GPa
Indentation modulus	72.5 GPa
Strength along the x-direction	6.5 GPa
Strength along the y-direction	6.6 GPa
Strength along the z-direction	3.1 GPa

4.4 Molecular Structure of C–S–H Model

A reasonable C–S–H model should not only be consistent with the experimental finding, but the local structures of the atoms in the molecular model are supposed to comply with the basic physical and chemical principles. In the following section, it is necessary to further analyze the bond and connectivity of the poor layered crystal. To better understand the poorly crystalized structure, the tobermorite 11 Å (Hamid model) is utilized for comparison.

4.4.1 Layered Structure

The simulated tobermorite 11 Å crystal and the C–S–H gel samples are shown in Fig. 4.4a, b. Correspondingly, the intensity profiles of different atoms at equilibrium states are plotted in Fig. 4.4c, d versus the distance in the z-direction. It can be clearly observed in Fig. 4.4a, b that calcium atoms and surrounding oxygen atoms form a Ca–O octahedral, constructing the Ca sheet; infinite long and defective silicate chains graft on both side of the Ca sheets in the tobermorite and C–S–H gel, respectively;

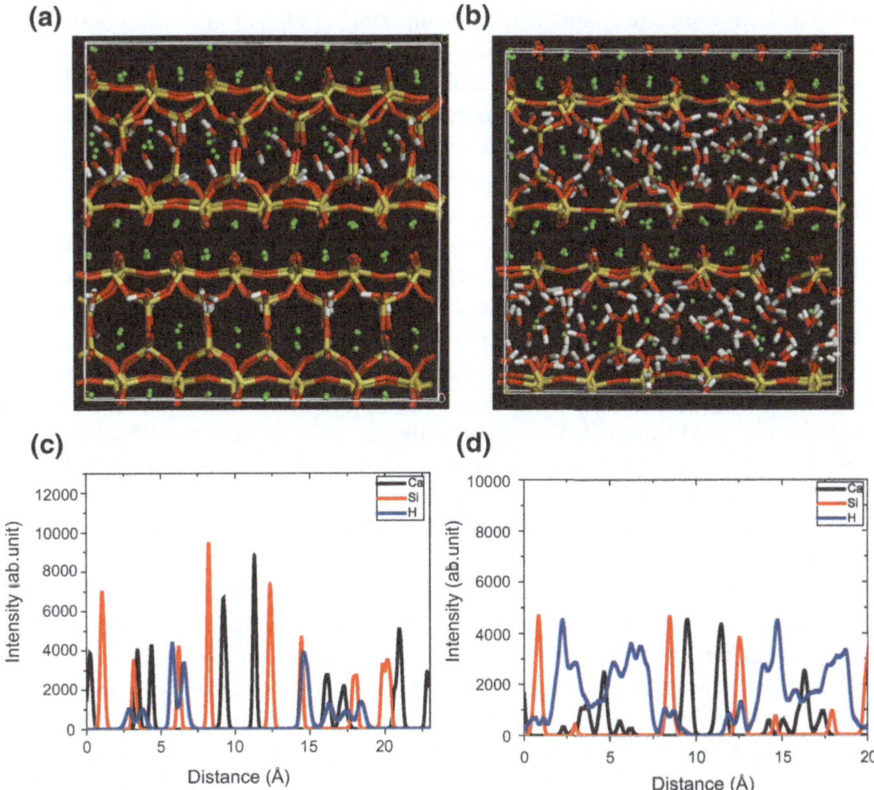

Fig. 4.4 Molecular structure of **a** tobermorite and **b** C–S–H gel; **c** atomic density distribution of Ca, Si, H atoms in the tobermorite 11 Å and **d** the C–S–H gel

between the neighboring calcium silicate sheets locate the interlayer calcium atoms (Ca_w), water molecules and also hydroxyl bonds. As shown in Fig. 4.4c, d, the alternative maxima of Ca, Si, and H atoms in density profiles indicate that C–S–H gel has a sandwich-like structure. The intensity peak value of Si in the interlayer region of C–S–H gel is sharply decreased as compared with that in the tobermorite, which is attributed to the removal of bridging silicate tetrahedrons in the amorphous phase. The missing silicate tetrahedrons result in disordered arrangements of the interlayer calcium structure and water molecules. On the one hand, the double intensity peaks of Ca_w atoms in the tobermorite are transformed to multi-peaks distribution continuous as through the interlayer region. It can be clearly observed in Fig. 4.4b that some Ca_w atoms diffuse into the defective region of the silicate chain and form a Ca_w–O bond with non-bridging oxygen (O_{NB}) atoms. On the other hand, the distribution of H atoms in the interlayer region implies that the structure of the water molecules, highly confined by the nanopore, is significantly distorted by the complicated local calcium silicate environment. The large intensity peaks of the H atoms also reflect that the defective silicate chains with large amounts of O_{NB} atoms have good ability to adsorb water molecules. Additionally, the intensity distributions of H atoms and Si atoms overlap, which means that water molecules not only are present in the interlayer regions but also diffuse into the defective regions of the calcium silicate sheet, as shown in Fig. 4.4c. It is also interesting to note that the silicate hydroxyl can be formed in the calcium silicate sheet, which enhances the association between water molecules and the defective silicate chains.

The smallest gel pores formed between neighboring calcium layers are widely distributed in the C–S–H gel and influence the physical movement and chemical reactivity of water molecules. It is necessary to characterize the porous structure of C–S–H gel. The pore volume and pore size distribution in both C–S–H gel and tobermorite crystal have been measured by using "Connolly surface" method [24] that is an accurate algorithm for nanopores calculation. The characteristic van der Waals radii of Si, Ca, and O atoms in the calcium silicate skeleton are 2.1, 1.97, and 1.52 Å respectively. The calculated porosity for the C–S–H gel is 26.5% that is around twice higher than that of tobermorite (14.2%). The large pore volume discrepancy between silicate structures can be explained by the atomic arrangement in the interlayer region. As shown in Fig. 4.5a, the bridging silicate tetrahedrons protrude into the interlayer region and connect with the calcium atoms by ionic-covalent bonds. The bridging SiO sites become the obstacles in the channel and separate the interlayer space into small pockets. On the other hand, Fig. 4.5b demonstrates that with removing of the bridging tetrahedrons, the connectivity of the interlayer channel is improved and small pockets are interconnected with each other. Furthermore, it can be observed from pore size distribution in Fig. 4.5c that the gel pores with size less than 3 Å occupy more than half volume in the crystal. It means that the water molecules with characteristic diameter 2.8 Å cannot access many small pockets in the interlayer region of tobermorite. For the C–S–H gel, the distribution curve shifts toward higher pore size and the pores with diameter ranging from 5 to 8 Å occupy predominated percentage. It contributes to the diffusive behavior of double water molecule layers. Additionally, the calculated porosity and critical pore length for the C–S–H gel matches well with previous computational results by Manzano et al. (porosity: 24%, pore length ~7.5 Å) [16], validating the accuracy for the computation method.

In order to give more quantitative insights into the structure of C–S–H gel and tobermorite, the local structure of Si, Ca atoms, and water molecules are further investigated by the radial distribution function (RDF) and the angle distribution.

4.4.2 Local Structure of Silicon

RDF curves of the Si–O bonds in the tobermorite and C–S–H gel are plotted in Fig. 4.6a. For the bonded atoms, the relevant bond distances can be readily determined by the positions of the peaks in the corresponding RDF. In the tobermorite and the C–S–H gel, the reactive force field yielded a Si–O bond distance of 1.620 Å, which is similar to the value of 1.615 Å obtained from previous simulations involving the BKS

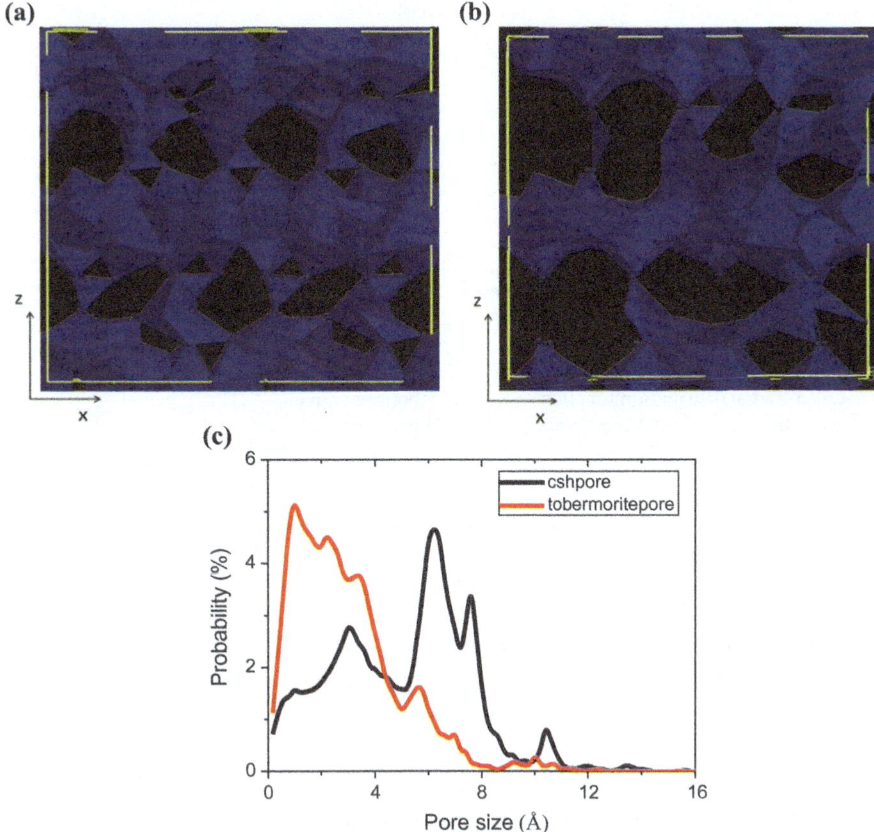

Fig. 4.5 Pore structure of the **a** tobermorite crystal; **b** CSH gel, the blue polyhedrons represent the calcium silicate skeleton; the dark space represent nanopores; **c** pore size distribution for tobermorite and CSH gel

model for the silicate glass [25, 26]. In the medium range, the Si–O spatial correlation in the tobermorite is extended to a larger distance than that for the C–S–H gel, in which the intensity peaks disappear around 5 Å. It implies an ordered arrangement of infinite long chain in the crystal phase. However, due to the removal of bridging silicate tetrahedron, the neighboring dimmer structures are separated from each other. The shorter distance correlation in the C–S–H gel can be interpreted as a weak connection between defective silicate chains.

The tetrahedral SiO_4 structures are typical in silicate-based systems, and can be depicted by the angle O–Si–O and Si–O–Si. As shown in Fig. 4.6b, for the O–Si–O angle, the reactive force field model yield peaks value at approximately 107.8° and 108.2°, and a full width at the half maximum (FWHM) of approximately 15° and 16° for the tobermorite and C–S–H gel, respectively. The experimental measurement of the O–Si–O angles using neutron diffraction gave a value of 109.7° and a FWHM

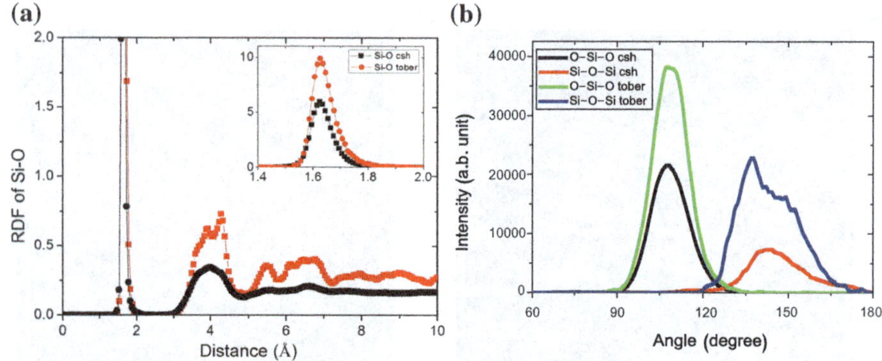

Fig. 4.6 **a** Radial distribution function of Si–O in tobermorite and C–S–H gel; **b** angle distribution of O–Si–O and Si–O–Si in tobermorite and C–S–H gel

of 10.6° [27]. The reactive force field can satisfactorily describe the silicon oxygen tetrahedral structures, since the simulated O–Si–O angle and FWHM are in reasonable agreement with previous observations. Additionally, the major difference for the angle distribution is the enlargement of the Si–O–Si angle in the C–S–H gel, which can be interpreted as the stretching of the neighboring silicon tetrahedrons. This is the consequence of the missing of the bridging silicate tetrahedron that constrains the extension of the dimmers along the y-direction.

4.4.3 Local Structure of Calcium Atoms

Figure 4.7a shows that the average Ca–O bonds obtained from the RDF in the tobermorite and C–S–H gel are 2.42 Å and 2.46 Å, respectively. The bond distance discrepancy mainly results from the different local environments of the calcium atoms. As shown in Fig. 4.7b, the coordination number (CN) of the calcium atom, describing the number of the neighboring oxygen atoms, ranges from 5 to 8, with 6 and 7 occupying dominant percentages. In respect of the chemistry crystals, the bond length and CN number of Ca–O are consistent with many experimental findings for the calcium silicate phases [28]. On average, the CN values of Ca for the tobermorite and the C–S–H gel are 6.7 and 6.8 respectively. The slight larger CN value for the C–S–H gel is mainly attributed to oxygen atoms from the water molecules that are abundantly distributed in the interlayer region. Hence, in the C–S–H gel, columbic attraction between Ca atoms and the associated oxygen atoms is relatively weaker than that in the tobermorite and the bond distance in the C–S–H gel is slightly larger.

The O–Ca–O angle distribution, shown in Fig. 4.7c, provides further information on the calcium-oxide local structure. The angle distribution for the tobermorite is characterized by three intense and narrow peaks at 60°, 73°, and 108°, and two small and broad peaks at 150° and 173° that all correspond to an ordered Ca–O

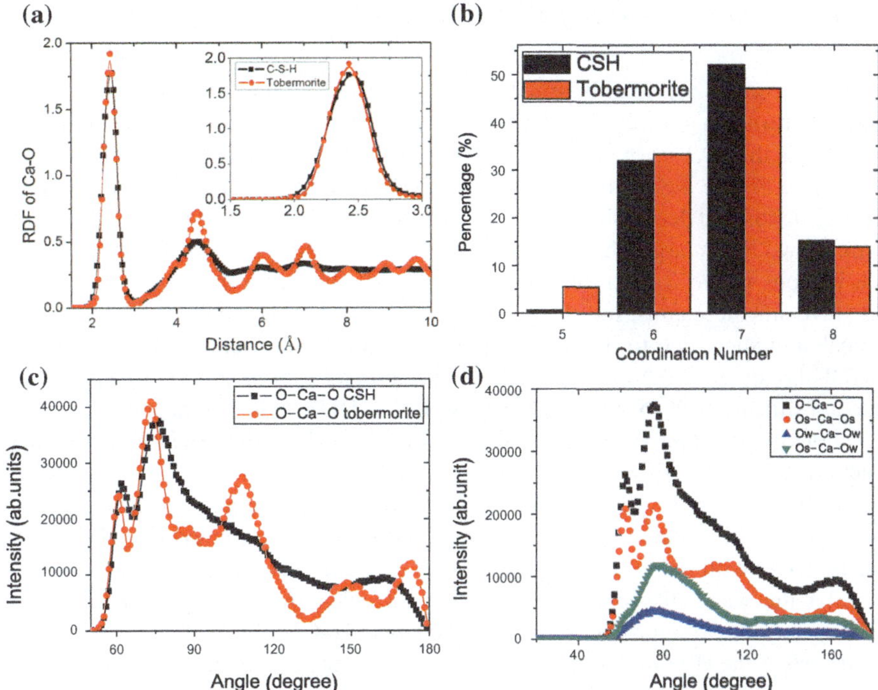

Fig. 4.7 **a** RDF of Ca–O in the tobermorite and C–S–H gel; **b** Ca–O CN distribution for tobermorite and the C–S–H gel, with a cutoff of 3 Å; **c** angle distribution of O–Ca–O for the tobermorite and C–S–H gel; **d** angle distribution of O–Ca–O in the C–S–H gel

octahedral arrangement. On the other hand, in the C–S–H gel, the peaks at 60° and 73° shift toward larger values, yet the intense peak at 108° disappears and the double peaks larger than 150° merge together. The deviation from the ordered octahedral is partly attributed to the distorted calcium silicate sheet with defective chains and partly caused by the calcium atoms that are strongly solvated by the interlayer water molecules. Figure 4.7d demonstrates O_w (oxygen atoms from water) and O_s (oxygen from silicate chains) components for the angle distribution of the C–S–H gel. The main contributions to the first peak at 60° are the O_s–Ca–O_s distribution and the angle is formed by one Ca atom with two O atoms in the same silicate tetrahedron. Even though both the O_s–Ca–O_s and O_w–Ca–O_w contribution make up the peak at around 73°, the sharp peak of the former and the broad one of the latter implies a more distorted octahedral structure in the interlayer region as compared with the calcium silicate sheet. It should be noted that the bond angle distribution can be taken as an essential feature of the poor crystal phase for the C–S–H gel. In the $CaSiO_3$, calcium silicate glass, double peaks at 60° and 90° can also be observed in the O–Ca–O distribution [26], implying the glassy nature for the C–S–H gel.

4.4.4   Local Structure of Water Molecule

Water molecules, confined between the neighboring calcium silicate sheets, dissociate and form Ca–OH and Si–OH bonds. The hydrolytic reaction rate is quite fast; during less than 0.1 ns, more than 44.3% of the water molecules in the C–S–H gel dissociate. Figure 4.8a shows the progressively increasing number of the Si–OH, Ca–OH bonds and decreasing of the water molecules. The ratio of the Ca–OH bonds is about two times larger than that of Si–OH bonds in the C–S–H gel. In the C–S–H gel, H^+ ions in the water molecules not only associate with the O_{NB} atoms in the defective silicate chains, but they also connect with the dangling oxygen atoms in the calcium sheet due to the elimination of the dimmer structures. In this respect, Ca–OH bonds can both be observed in the interlayer region and the surface of the calcium silicate sheet so that the percentage of Ca–OH bonds significantly increases. Additionally, the number of hydroxyl bonds in the C–S–H gel grows smoothly and reacts faster. The high reactivity for the C–S–H gel is due to the richness of the O_{NB} (non-bridging oxygen) atoms in the defective silicate chains that accelerate the dissociation of the water molecules. In fact, during the reaction process, H^+ ions do not associate with the bridging oxygen atoms that exhibit hydrophobic nature [16].

After the water dissociation, the water molecules are strongly connected with the neighboring structure by forming Si–OH and Ca–OH bonds. More importantly, the nature of the H-bonds is also transformed from interaction between neighboring water molecules to connection between the hydroxyl groups and the oxygen atoms from the silicate chains. The oxygen atoms are categorized into O_h (oxygen atoms in the hydroxyls), O_s (oxygen atom without H connection) and O_w. The H-bonds connections for the three types of oxygen atoms are listed in Table 4.3.

On average, each water molecule has 2.42 H-bonds connections with neighboring water molecules and calcium silicate sheets, which are close to 2.43 H-bonds calculated in a previous simulation by the empirical force field CSHFF [9]. However, in the current simulation, the confined water can accept some H-bonds from the oxygen atoms on the silicate chains, because the O_{NB} atoms are hydroxylated. In the case of simulation by CSHFF, water molecules are not allowed to dissociate so that the O_{NB} atoms in the silicate chains cannot form H-bonds to the confined water molecules. In regard to the H-bonds distribution, the reactive force field describes a more reasonable structure than the empirical force field. Furthermore, each hydroxyl group can donate 1.13 to and accept 1.18 H-bonds from both the local water molecules and the calcium silicate sheets. Considering that more than 44.3% of the total water molecules are transformed to hydroxyl groups, the H-bonds contributed by O_h–O_h and O_h–O_w form the dominant ratio in the total H-bonds network. Because the bond strength of O_h–O_w is much larger than that of O_w–O_w [9], the mechanical properties of the C–S–H gel can be further improved after the hydrolytic reaction. Additionally, each un-hydroxylated oxygen atoms in the calcium silicate sheet can accept 0.32 H-bonds from the surrounding hydroxyl groups and water molecules. It implies that only a few, weak H-bonds are contributed by the bridging oxygen atoms.

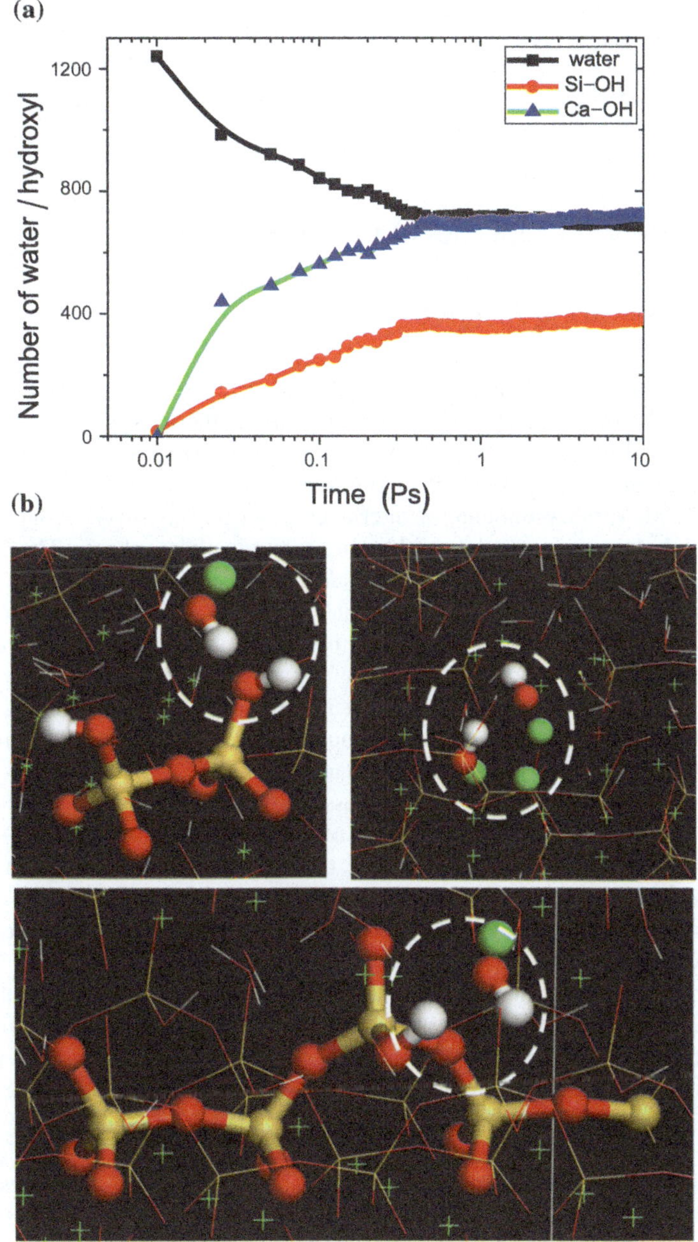

Fig. 4.8 **a** Evolution of water and hydroxyl numbers in the C–S–H gel. **b** Snapshots of water dissociation in the C–S–H gel

Table 4.3 Average number of hydrogen bonds per O_s, per hydroxyl and per water molecule in the C–S–H gel

	Acceptor	Donator
O_s	0.32	0.00
O_h	1.18	1.13
O_w	1.29	1.13

4.5 Mechanical Properties of C–S–H Gel

4.5.1 Stress–Strain Relations

Previous structural and dynamic analyses provide a clear picture of chemical bonds in two-layered structures. The silicate chains provide the most stable backbone of C–S–H gels and tobermorite. Ca atoms are associated with the neighboring O in the silicate chains and develop high strength calcium silicate sheet. The Ca_w atoms associated with various amounts of interlayer water and hydroxyl groups play an important role in connecting the neighboring calcium silicate sheets. The mechanical properties of the layered structures are determined by the combination of these chemical bonds and uniaxial tension testing.

The stress–strain curve can characterize the mechanical behavior of layered structure during the tensile process and help gain insights into the constitutive relation between stress and strain. Different stress–strain relations of tensile loading in the x, y, and z-directions indicate the heterogeneous nature of the layered structures.

The relations for C–S–H gel are first analyzed. In the x-direction, as shown in Fig. 4.9a, during the tensile process, stress first increases linearly in the elastic stage and subsequently slowly increase to a maximum value of 6.5 GPa at a strain around 0.17 Å/Å. After the maximum value, the stress directly drops without obvious yield behavior and the stress slowly decreases to zero as the strain reaches 0.8 Å/Å. Stress–strain curves in the x and y-directions are almost coincident with strain from 0 to 0.5 Å/Å, but the stress in the y-direction slowly reduces from 0.3 to 0.8. The structure of C–S–H gel cannot draw fracture along the y-direction even as the strain reaches 0.8 Å/Å, implying good plasticity. The deviation illustrates the different deformation mechanism on the XY plane. The C–S–H gel is more likely to break in the z-direction and the strain at the fracture state is 0.4 Å/Å, indicating that the interlayer structure has a more brittle feature.

In regard to the tobermorite, the major feature of the stress–strain relation is the mechanical performance in the y-direction. As shown in Fig. 4.9b, the stress–strain curve is quite different than that in the x- and z-directions: the stress jumps quickly to 18 GPa in less than 0.1 Å/Å, then after a small drop at strain 0.13 Å/Å, the stress further climbs to 19.5 GPa and finally decreases as the strain reaches 0.33 Å/Å. The high failure strength and stress enhancement in the late stage implies that Si–O bonds, growing along the y-direction, are better in load carrying. It should be noted that the Young's modulus, the tensile strength and the toughness along the y-direction are reduced to a great extent as the infinite long silicate chains in tobermorite crystal

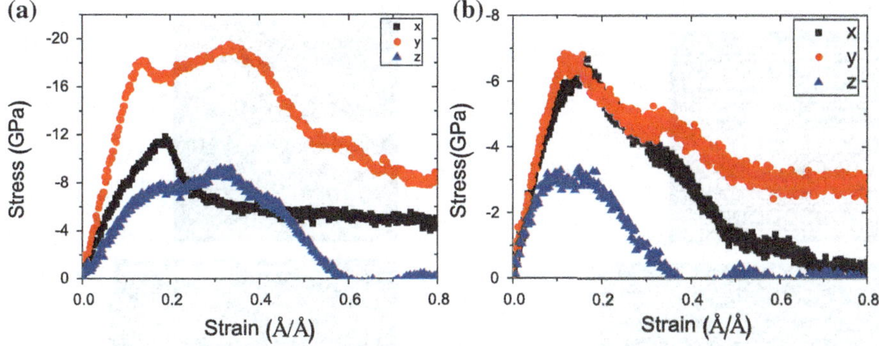

Fig. 4.9 Stress–strain relation of layered structures tensioned along x, y and z-direction of **a** C–S–H gel; **b** tobermorite

are transformed to isolated dimmer structures in the C–S–H gel. The mechanical properties discrepancy between tobermorite and the C–S–H gel confirms that the silicate chain, the skeleton of the calcium silicate phases, plays an essential role in loading resistance. In addition, as compared with the strength in the x and z-directions in the C–S–H gel, the strength of the tobermorite is much larger, also implying mechanical contribution from the bridging silicate tetrahedron. Based on the idea of increasing the silicate chain length, some biomimetic approaches, such as grafting organic moieties on the C–S–H mineral lamellae, have been proposed to improve the mechanical properties of cementitious system [29].

4.5.2 Chemical Reaction in the Deformed C–S–H Gel

The configurations of the Ca, Si, O and H atoms from strain = 0 Å/Å to strain = 0.6 Å/Å are listed in Fig. 4.10a, b for preliminary qualitative illustration of the damage process of tobermorite and the C–S–H gel, respectively. For the tobermorite, the siloxane bonds grafted in the calcium sheet, subjected to tensile loading, are elongated in the elastic region. The Si–O–Si angle is stretched open to take up the strain in the silica system. In the yield region, there is a gradual breakage of the siloxane bonds that allows transformation of the morphology. It is seen in Fig. 4.10a that at strain 0.4 Å/Å, the layered crystal phase gradually changes to an amorphous one due to bond breakage and reconnection. More importantly, it can be observed that some distorted silicate chains are twined together, which enhances the interlayer connections. In the tensile period, the small cracks are created but the local structure rearrangement slows down the crack propagation, as shown in the configuration at strain 0.6 Å/Å. On the other hand, deformation of the C–S–H gel demonstrates a different failure mechanism. As shown in Fig. 4.10b, at a strain 0.2 Å/Å, the Ca–O bonds elongation and distortion of the calcium octahedron take up the major strain at

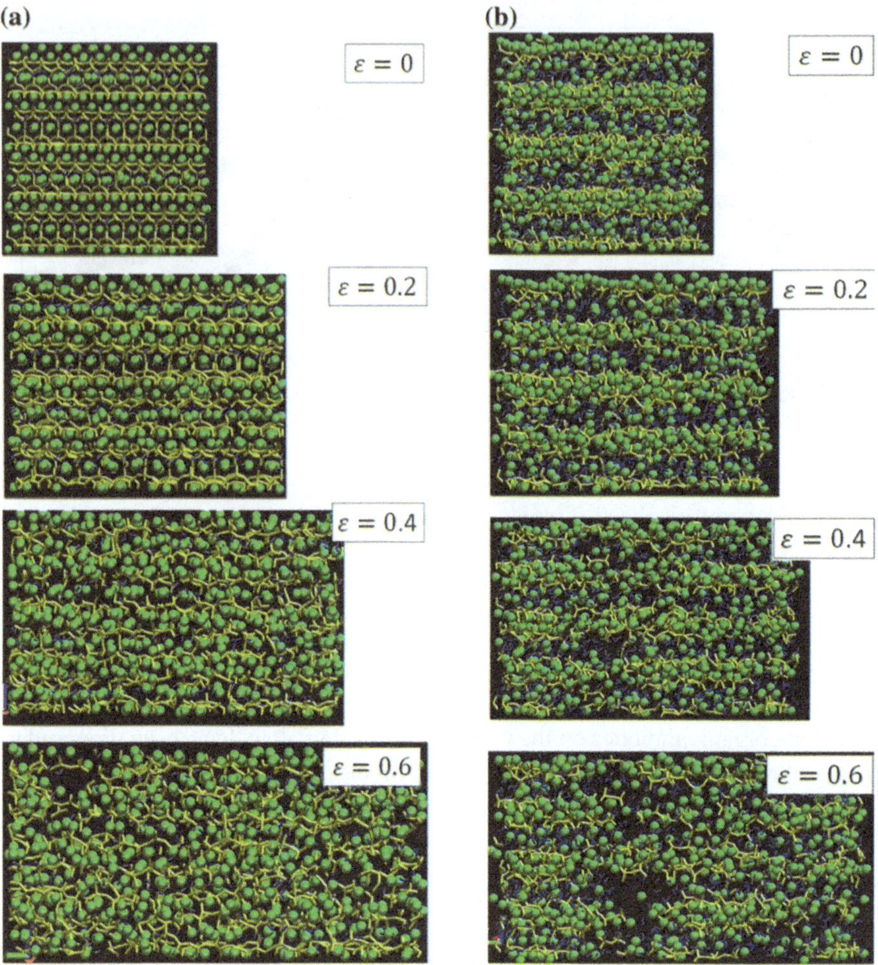

Fig. 4.10 Molecular structure evolution of **a** tobermorite; **b** C–S–H gel tensioned along y-direction. The yellow sticks are the silicate chains; green balls are calcium atoms; blue lines are the water molecules

the initial stage. Because the dimmer structures are distributed separately, the silicate chains cannot provide sufficient mechanical contribution by stretching the Si–O–Si bonds. At the failure stage, the cracks grow and coalesce rapidly through the region with defective silicate chains as shown in Fig. 4.10b which describes the strain states from 0.4 to 0.6 Å/Å, resulting in a continuous decreasing of the stress.

In order to investigate the morphological evolution quantitatively, the change of Q_n percentages is calculated to describe the structural variation of the silicate skeleton during the tensile process, as shown in Fig. 4.11.

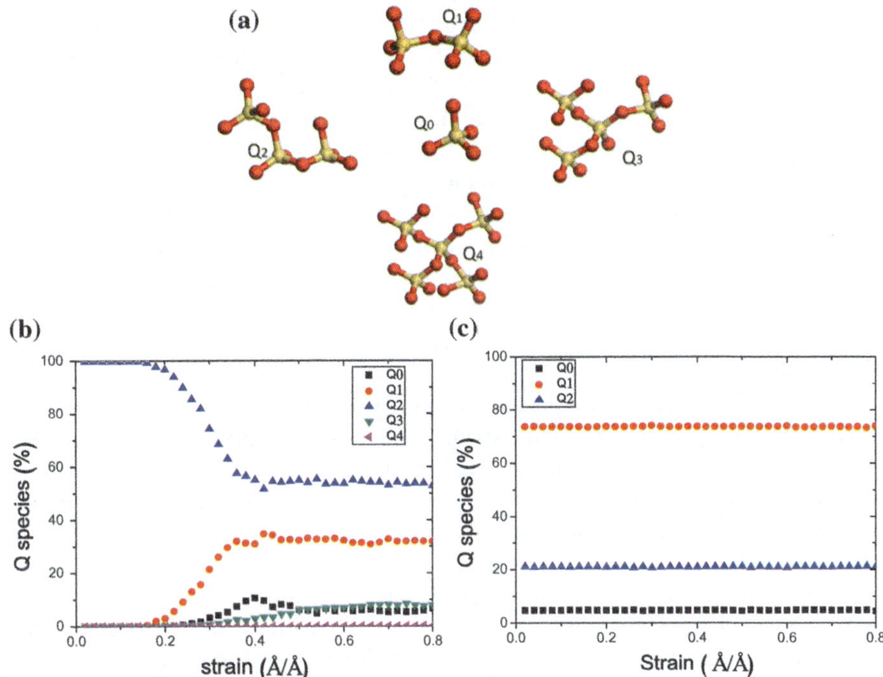

Fig. 4.11 **a** Molecular structure of silicate tetrahedron Q_n ($n = 0, 1, 2, 3, 4$); **b** evolution of the connectivity factor Q species of the tobermorite tensioned along y-direction; **c** Q species evolution of the C–S–H gel tensioned along y-direction

As in the silicate composite structure, the connectivity factor, Q_n ($n = 0, 1, 2, 3, 4$) is an important parameter in estimating the silicate connection, where n is defined as the number of connected neighboring silicate tetrahedrons. As shown in Fig. 4.11a, Q_0 is the monomer; Q_1 represents the dimmer structure (two connected silicate tetrahedrons); Q_2 is the long chain; Q_3 is the branch structure; Q_4 is the network structure [30]. The infinite long silicate chain in the tobermorite varies greatly during the tension process. As shown in Fig. 4.11b, the Q species evolution can be clearly distinguished in three stages as the strain increases along the y-direction. Initially, the percentage of Q species maintains unchanged during the first 0.16 Å/Å strain period, in which stage, the silicate chain length is elongated and Si–O–Si angle is stretched open to carry the loading. Subsequently, when the strain exceeds 0.16 Å/Å, Q_2 begins to decrease, and Q_1 and Q_0 increases. It means that some parts of the silicate chains are stretched broken, resulting in the morphological transformation from long silicate chains to dimmer structures. Meanwhile, the Q_3 species grows at strains of 0.3 Å/Å, implying formation of the branches structures. Some silicate chains that were stretched broken can reconnect to form a new silicate skeleton. The newly connected branches can bridge the neighboring calcium silicate layers and construct the threedimensional structure. It is worth noting that the structural rearrangement significantly improves the mechanical performance. By previous ab initio calculation [31], the presence of the Q_3 species can improve the interlayer stiffness of tobermorite (a kind of calcium silicate crystal) [4] by forming a hinge mechanism. It explains why the stress continues to increase after the initial drop in the stress–strain relation curve in Fig. 4.11c. Finally, the ratios of different Q species remain unchanged, as the strain reaches 0.4 Å/Å, from which the stress continues to reduce. However, the Q species does not change during tension along the y-direction for the C–S–H gel. It implies that defective silicate chains, isolatedly distributing in the C–S–H gel, have not been drawn broken. Due to the missing bridging silicate tetrahedrons, the mechanical contribution from the Si–O bonds is significantly reduced. Hence, the tensile strength and stiffness of the C–S–H gel in the y-direction is weakened to a great extent.

Furthermore, it should be noted that in Fig. 4.8b, the intrusive water molecules attack the Si–O–Si bonds, and many Si–OH groups and Ca–OH bonds are formed at the end of the crack tips, which further weakens the calcium silica structure in loading resistance. Hence, the number of water molecules and Si–OH and Ca–OH groups was recorded during the tensile process to quantitatively analyze water dissociation process.

As shown in Fig. 4.12a, b, the number of water molecules continuously decrease during the tensile process, implying that hydrolytic reaction widely happens. In the C–S–H gel, the number of Si–OH and Ca–OH both increase simultaneously and the increased amount is the same. The reaction process is that water molecules dissociate into H^+ and OH^-, the H^+ ions diffuse to associate with the O_{NB} atoms in the silicate chains, while the OH^- ions form Ca–OH bonds with neighboring interlayer calcium atoms. On the other hand, the water dissociation mechanism in the tobermorite is quite different. As shown in Fig. 4.12b, during the tensile process, while the number of Si–OH bonds continuously increases, the number of Ca–OH bonds first increases and then slightly decreases. It means that the dissociated water molecules mainly react with the broken silicate chains rather than the Ca atoms.

The reaction mechanism can be categorized into two types: the first one is caused by chemical adsorption of the water and the second one is caused by Si–O–Si bond breakage. The chemical adsorption for the water molecules follows the reaction pathway in the Fig. 4.12c. Water diffuses and is adsorbed near the neighboring silicon atoms; bond formation between the water and silicon atoms, fivefold silicate structure forms; water dissociation; proton transfers, and $Si–O_B$ bond breakage. In the dissociation process, the water molecules lead to the formation of a penta-coordinated silicon structure, which is energetically unfavorable. The unstable structure subsequently relaxes and dissociates with the O_B atom. In this way, the water molecules attack the tensile siloxane bonds and accelerate the separation between the neighboring Q_4 species. Interestingly, the reaction is reversible for the oligomerization of two silicate monomers [31]:

$$Si(OH)_4 + Si(OH)_4 \rightarrow (HO)_3SiOSi(OH)_3 + H_2O$$

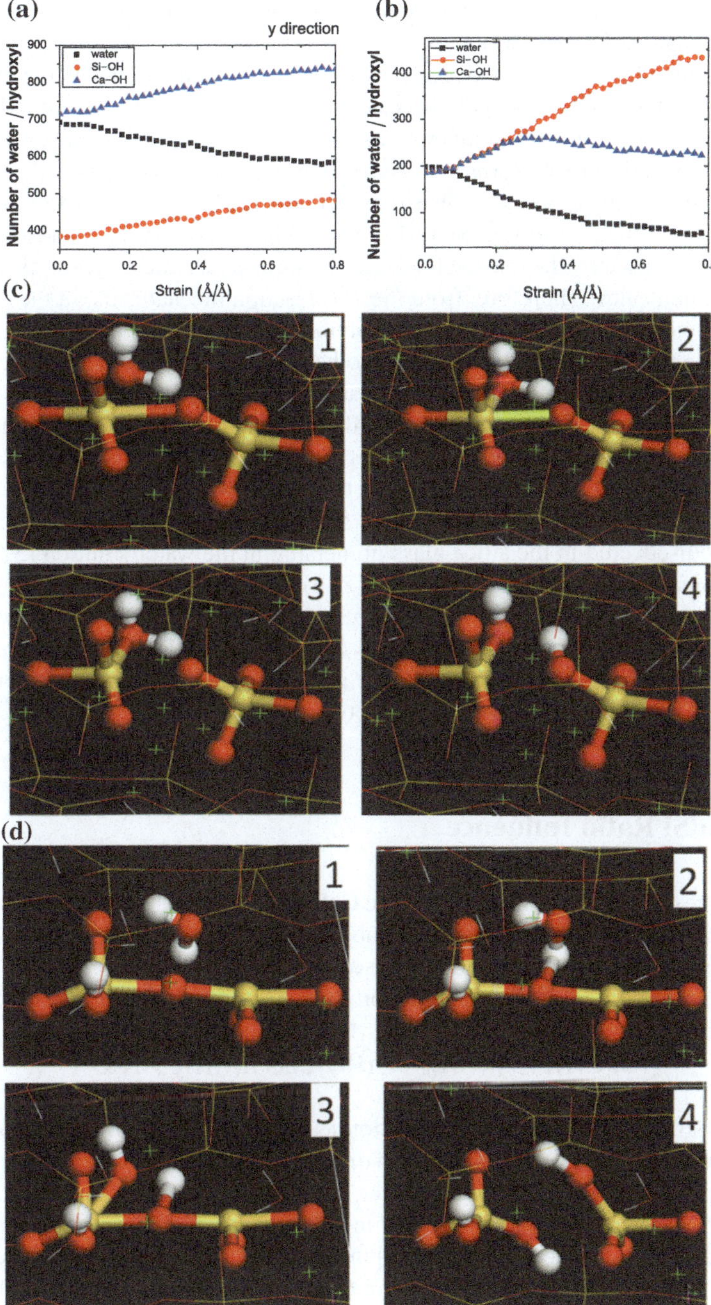

Fig. 4.12 The number of water molecules, Si–OH bonds and Ca–OH bonds evolution with strain in **a** the C–S–H gel; **b** the tobermorite; **c** reaction pathway of chemical adsorption; **d** reaction pathway of Si–O–Si breakage

monomers de-protonate; ionized monomers form penta-coordinate silicon; unstable bond breakage and water formation.

On the other hand, the second reaction occurs later as compared with the first reaction at the high strain level stage. The second reaction mechanism of water dissociation is listed in the following sequence, as shown in Fig. 4.12d: the siloxane bond is initially stretched and SiO_4 turns to disordered; water adsorbs with O_B in the Si–O–Si bond; water dissociates to the hydroxyl group; the free dissociated OH group bonds with the other silicon, forming Si–OH. It should be noted that the distorted silicate tetrahedron, caused by the tension loading, in return, accelerate a hydrolytic reaction for water molecules. Therefore, from the two reaction mechanisms, in the presence of tensile loading, the hydrolytic and depolymerization reaction are interplayed and enhance the reaction degree between each other. Additionally, despite different reaction pathways, the same reaction products are separated into two Si–OH groups. The newly produced Si–OH structures are hard to polymerize again because they have to overcome the energy barrier of oligomerization.

In this way, the water molecules attack the tensile siloxane bonds and accelerate the separation between the neighboring Q_2 species. The "hydrolytic weakening" has also been discovered in the silica glass immersing in the water solution [32]. Thermodynamically, according to the theory of Zhu et al. [32], increasing the stress in the C–S–H and tobermorite can reduce the energy barrier for activating the hydrolytic reaction, which is the fundamental reason for water dissociation. Therefore, in the presence of tensile loading, the hydrolytic and depolymerization reaction are interplayed and enhance the reaction degree between each other.

4.6 Ca/Si Ratio Influence

The silicate chains, as the skeletons of the C–S–H gel, change significantly with different Ca/Si ratio. The cementitious hydration, the production process of C–S–H gel, is based on the sol–gel reaction in the presence of calcium ions. The oligomerization of silicate monomers follows the water-producing condensation reaction [31]:

$$Si(OH)_4 + Si(OH)_4 \rightarrow (HO)_3SiOSi(OH)_3 + H_2O$$

As illustrated in Fig. 4.13a, the reaction leads to linkage of silicic monomers to form a dimmer structure and dissociation of one water molecule. The polymerization reaction produces different silicate cluster.

Both NMR studies [33, 34] discussed in Fig. 2.11 in Chap. 2 and sol–gel reaction simulated by the reactive force field [18] indicates that mean silicate chain length in the C–S–H gel decreases with increasing Ca/Si ratio. As shown in Fig. 4.14, with increasing calcium content, the C–S–H gel transforms from the network and branches morphology to the long silicate chains and finally degrades to short silicate chains. The structural transition is attributed to the depolymerization role of Ca atoms.

Fig. 4.13 Polymerization
reaction for the silicic
monomers

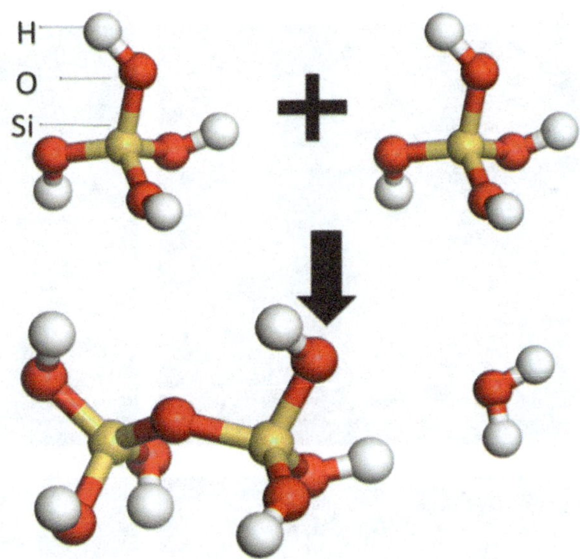

4.6.1 Model Construction at Different Ca/Si Ratios

In the C–S–H model, the silicate chains were broken to match well with the Q
species distribution obtained from the NMR test [34]. As shown in Fig. 4.15a, the
mean silicate chain lengths (MCL = 2(Q2/Q1 + 1)) of C–S–H with different Ca/Si
ratios are initially set according to experimental trend. Figure 4.15b schematically
illustrates that the elimination of the bridging silicate tetrahedron can both increase
Ca/Si ratio and decrease the mean silicate chain length. Ten C–S–H samples with
different silicate morphologies are constructed with Ca/Si ratio ranging from 1.1 to
2.0. The dry structures with Ca/Si ratio of 1.1, 1.4, 1.7 and 2.0 are plotted in Fig. 4.16a,
b, c and d, respectively. When the dry structures are obtained, the GCMC water
adsorption and MD equilibrium runs are followed to construct the C–S–H models
at different Ca/Si ratios. The latter two steps are same as those aforementioned in
Sect. 4.2.

4.6.2 Molecular Structures at Different Ca/Si Ratios

Four simulated C–S–H gel samples at different Ca/Si ratios are shown in Fig. 4.17.
Correspondingly, the intensity profiles of different atoms of 10,000 configurations at
equilibrium state are plotted in Fig. 4.18 versus the distance in z-direction. As shown
in Fig. 4.18, with increasing Ca/Si ratio, intensity peak value of Si in the interlayer
region gradually disappears (Ca/Si = 1.7 and Ca/Si = 2.0), due to the missing of
bridging silicate tetrahedron. In addition, because eliminating silicate chains disturbs

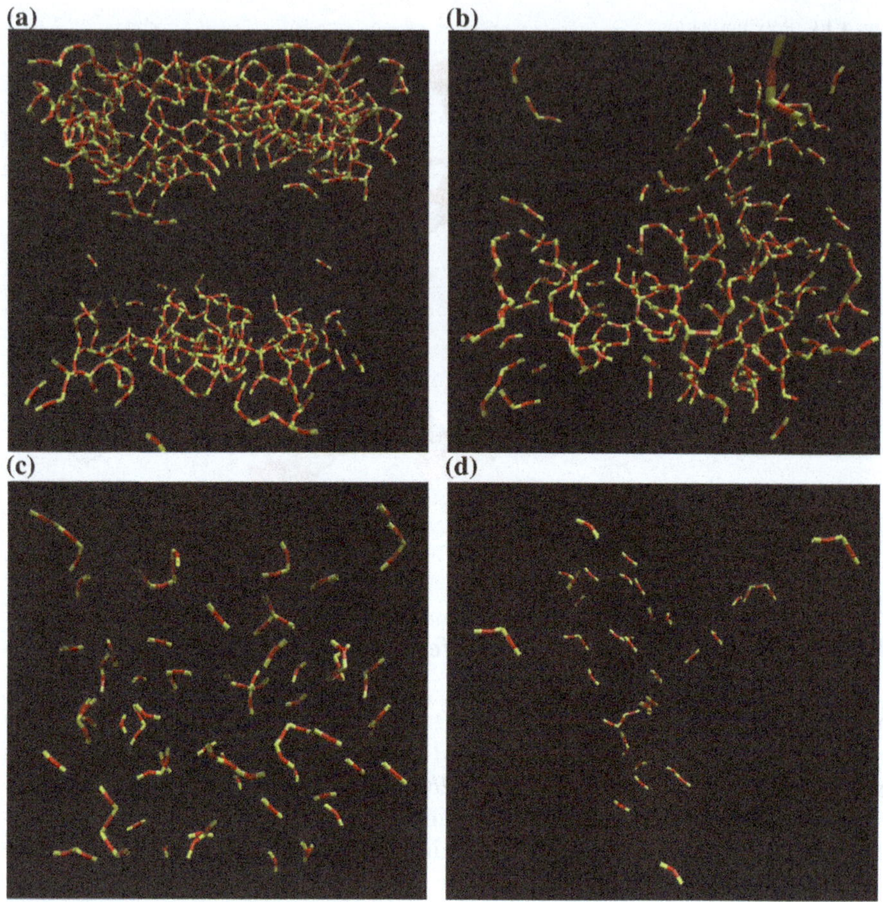

Fig. 4.14 Silicate cluster morphologies at different Ca/Si ratio

the calcium sheet arrangement and makes the atoms distribution more amorphous, double intensity peak of Ca_s is not that pronounced at Ca/Si = 2.0 as shown in Fig. 4.18d.

In order to quantitatively analyze the water content in the C–S–H gel, the ratios of H_2O/Si are calculated and plotted versus the Ca/Si ratios. Consistent with the result in the intensity profiles, while Ca/Si ratio changes from 1.1 to 2.0, H_2O/Si ratio increases from 0.67 to 2.36, implying the better water adsorption ability of C–S–H gel at high Ca/Si ratio. Meanwhile, water molecules, confined in the nanopores, dissociate into the hydroxyl groups. The dissociation degree, defined as the dissociation ratio of water molecules, can be used to estimate the reactivity of water in the C–S–H gel confinement. As shown in Fig. 4.19a, the OH/Si ratio increases from 0.2 to 1.2 and the dissociation degree also improves from 30.7 to 52%, implying the chemical activity of water molecules in the C–S–H gel with high Ca/Si ratio. Both the adsorption

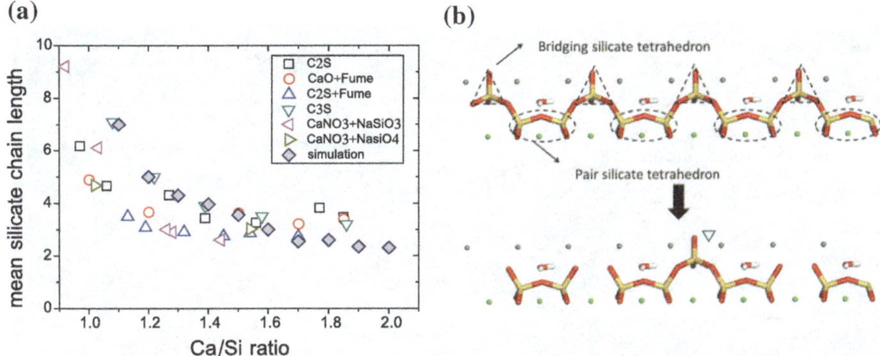

Fig. 4.15 **a** Silicate chain length variation with Ca/Si ratio. The experimental and computational data was achieved from Refs. [18, 34]. **b** Schematically illustration of the short silicate chain formation

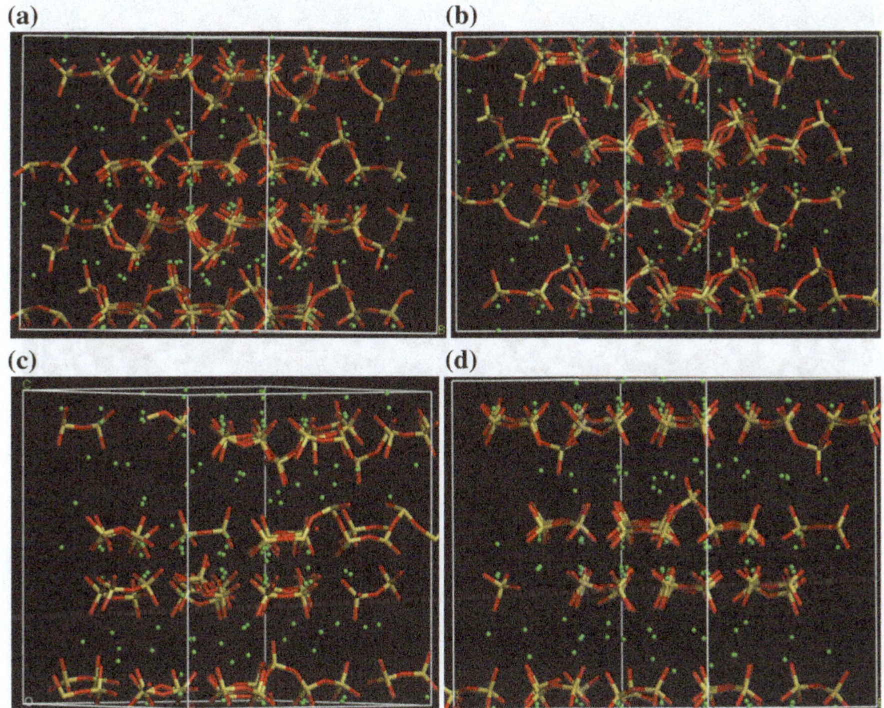

Fig. 4.16 Initial dry C–S–H structure of Ca/Si ratio **a** 1.1, **b** 1.4, **c** 1.7 and **d** 2.0. Simulation box size: a = 21.3 Å, b = 21.2 Å, c = 21.9 Å; $\alpha = 90°$, $\beta = 90°$, $\gamma = 90°$

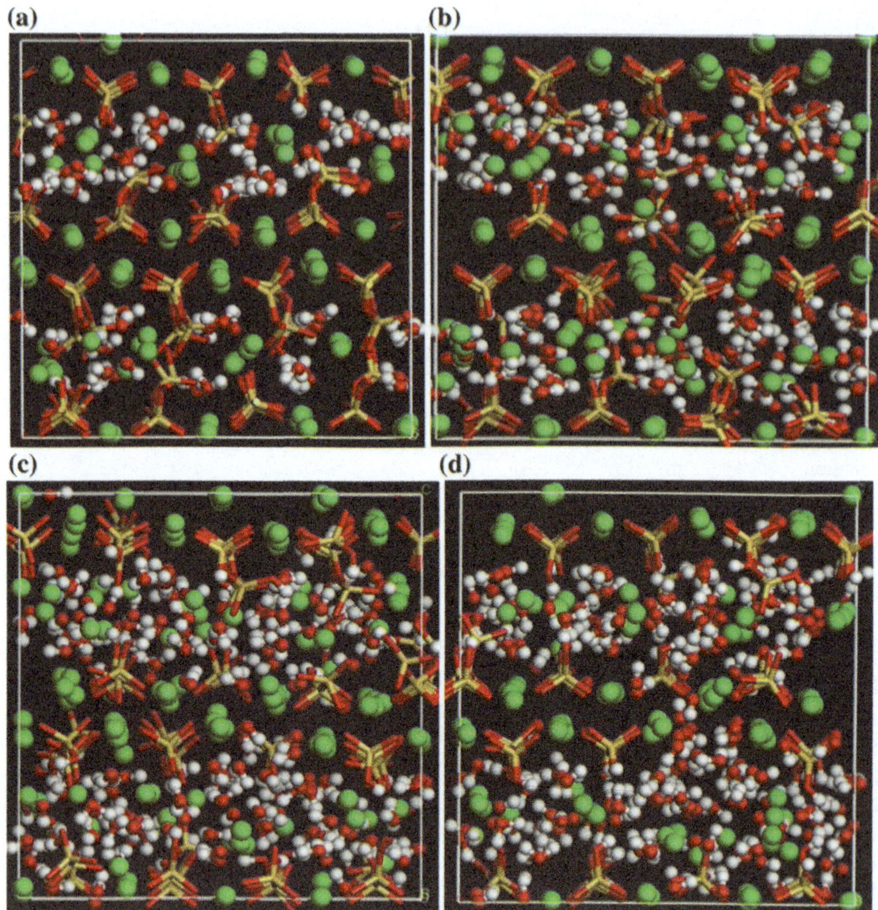

Fig. 4.17 Snapshots for the C–S–H samples at Ca/Si ratio **a** 1.1, **b** 1.4, **c** 1.7 and **d** 2.0 at equilibrium state

ability and reactivity of water molecules are closely related to the geometric and electronic nanopore environment. In particular, the O_{NB} atoms in the silicate chains are responsible for the water adsorption and dissociation. With increasing Ca/Si ratio, the silicate chain length gradually decreases and the more defective silicate chains contribute to larger amount of O_{NB} atoms. The hydrophilic nature of O_{NB} atoms leads to the water adsorption and dissociation. On the other hand, the defective silicate chains at high Ca/Si ratio have a smaller number of O_B atoms.

It is valuable noting that there are both Si–OH and Ca–OH in the simulated C–S–H gels with different Ca/Si ratios. Si–OH and Ca–OH bonding are important to determine the local structure of the C–S–H gel. As exhibited in Fig. 4.19b, the simulated Ca–OH/Ca ratio continues increasing with Ca/Si ratio, which is consistent with the trend obtained from the inelastic neutral scattering (IENS) [15]. However,

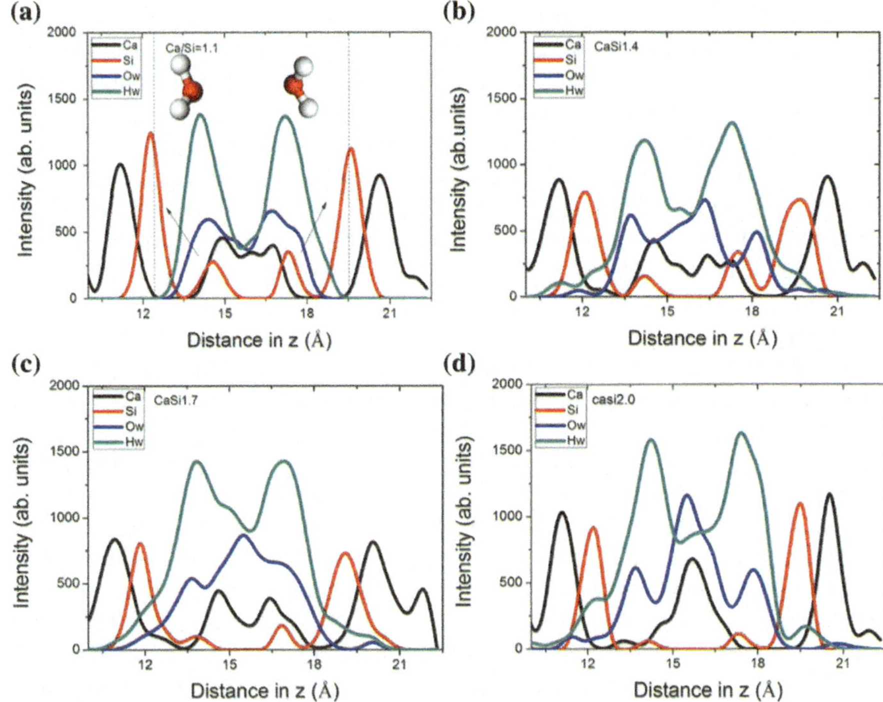

Fig. 4.18 Atomic intensity of the C–S–H sample at Ca/Si ratio of **a** 1.1, **b** 1.4, **c** 1.7 and **d** 2.0

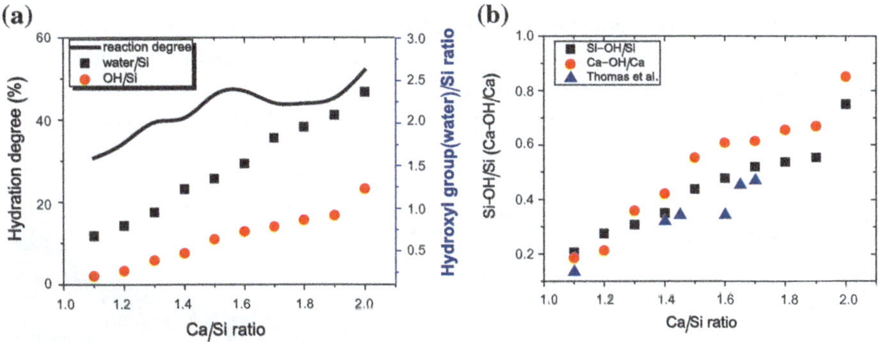

Fig. 4.19 **a** Number of water and hydroxyl group, and hydrolytic reaction degree vary with Ca/Si ratio; **b** number of Si–OH and Ca–OH change with Ca/Si ratio

the Ca–OH/Ca ratios from experiment are quite smaller than those from the current simulation. The H_2O/Si ratios of the C–S–H gel prepared for the IENS testing range from 0.95 to 1.34, which is quite smaller than the water content in our simulation. It might result in the lower percentage of Ca–OH bonds.

The hydrolytic reaction also has great influence on the dynamic properties of the C–S–H gels. Figure 4.20 exhibits that the mobility of atoms ranks in the following order: $O_s > O_h > O_w$. MSD characterizes the chemical bonding and physical associated water molecules, as mentioned in the C–S–H models proposed by Powers and Brownyard [35], and Feldman and Sereda [36]. The hydrolytic reaction transforms part of surface adsorbed water molecules into the "immobile" hydroxylation layers. In the transition layer, hydroxyl groups, resembling the dynamic nature of calcium silicate skeletons, are strongly restricted and cannot diffuse freely. Only vibration and rotation behavior of the hydroxyl bonds result in the low MSD values of the chemical bonding water.

As shown in Fig. 4.21, with increasing Ca/Si ratio, the average H-bond number gradually increases from 2.27 to 2.71, which is mainly attributed to the richness of water molecules in the C–S–H gel with high Ca/Si ratio. According to the accepting and donating role, and the O types, the H-bonds are further decomposed into five parts. On average, while the number of H-bonds donating to the O_s atoms progressively decreases from 1.28 to 0.33, the number of H-bonds donating to and accepted by the O_h atoms increase to 0.51 and 0.85, respectively. On one hand, the long silicate chains translate into short dimmer structures with missing of O_s atoms as the Ca/Si ratio increases. On the other hand, part of the O_w and O_s atoms become O_h atoms after the hydrolytic reaction, which widely happens at high Ca/Si ratio. Both effects result in increasing O_h number and decreasing O_s number and influence the corresponding H-bonds of O_w–O_s and O_w–O_h. Besides, the amount of O_w–O_w connections also increases from 0.266 to 0.516, which is mainly contributed by the interaction between neighboring molecules in the third layer as illustrated in the atomic profile. In the previous simulation by empirical force field CSHFF, 2.34 H-bonds connected

Fig. 4.20 Mean square displacement of O_s, O_w and O_h atoms

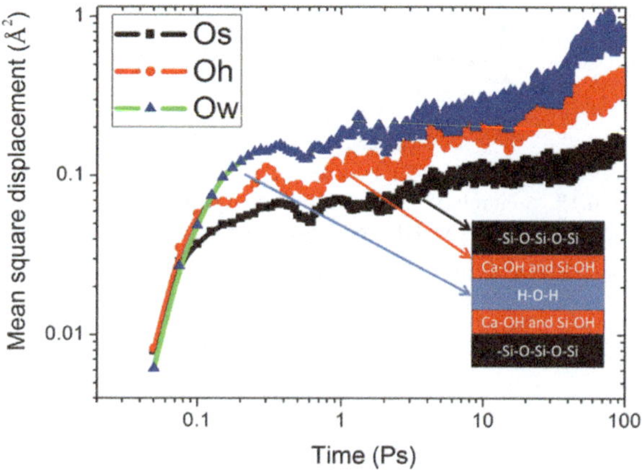

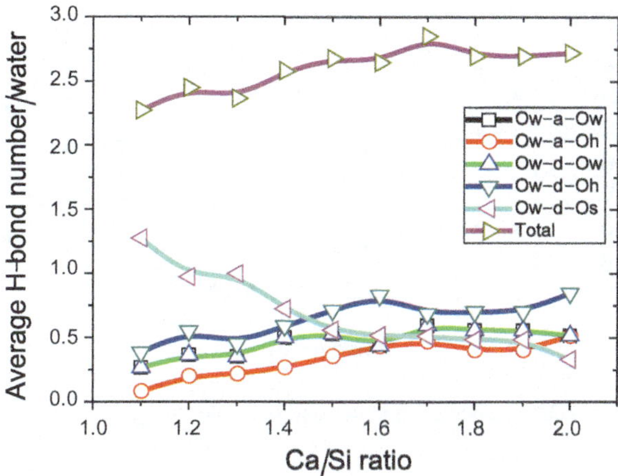

Fig. 4.21 Average H-bond number evolution with Ca/Si ranging from 1.1 to 2.0

by silicate chains and neighboring water molecules are calculated in Pellenq's model [9] by using Kumar et al. method [37]. At Ca/Si ratio 1.7, the H-bond intensity simulated by reactive force field increases about 21.7% as compared with that from empirical one. CSHFF cannot describe the water reactions, so no hydroxyl groups exist in the nanopores. However, since 44.7% of the water molecules dissociate into hydroxyl group and form the chemical bonds in current simulation, hydroxylation layer deeply embedded into calcium silicate sheets densified the H-bond network.

4.6.3 Mechanical Properties at Different Ca/Si Ratios

Modulus and Tensile Strength

Uniaxial tension test is applied to ten C–S–H samples in x-, y-, and z-direction to investigate the Ca/Si ratios influence on the mechanical properties of the material. The tensile strength and Young's modulus, calculated from the stress–strain curves, are plotted versus Ca/Si ratio in Fig. 4.22. All the C–S–H samples show the heterogeneous mechanical behavior: XY plane demonstrate larger cohesive force and stiffness than those along interlayer direction at all the Ca/Si ratios. As shown in Fig. 4.22b, in y-direction, with increasing Ca/Si ratios, the tensile strength gradually reduces from 14.5 to less than 7 GPa. In Fig. 4.22a, Young's modulus decreases from 113.7 to around 62.5 GPa as Ca/Si varies from 1.1 to 1.7 and maintains at around 62 GPa as Ca/Si from 1.7 to 2.0. The strength weakening trend can also be observed in the tensile samples along x-direction: tensile strength decreases from 13 to 5.7 GPa accompanied by Young's modulus reduction from 94 to 50 GPa. In regard to the properties along interlayer direction, as shown in Fig. 4.22, in z-direction, the tensile strength decreases from 6.68 to less than 2.7 GPa, with Young's modulus reducing

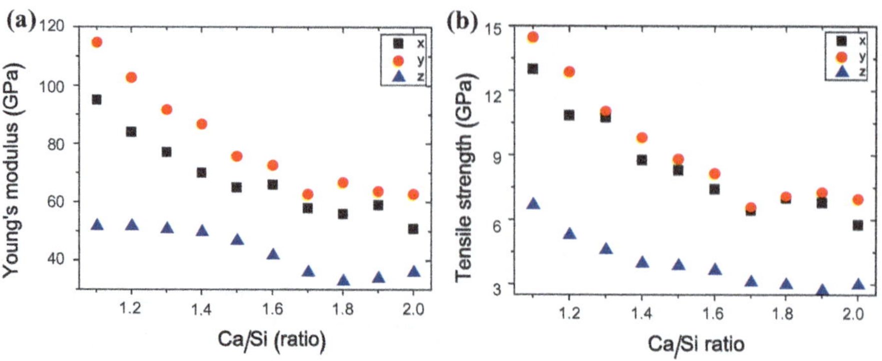

Fig. 4.22 **a** Young's modulus and **b** tensile strength evolution with Ca/Si ratio

from 52 to 33 GPa. As compared with that in *XY* plane, the weaker stiffness in *z*-directions are more approximate to the results from recent nanoindentation test on synthesized C–S–H gels [38]: when the Ca/Si molar ratio of C–S–H increases from 0.7 to 2.1, elastic modulus achieved from experiment reduces from 27 to 20 GPa. Because nanoindentation cannot eliminate the influence of nanoporosity that weakens the mechanical properties significantly, the experimental results demonstrate lower values than those obtained from our simulation.

The silicate morphology and the structure of interlayer water molecules are both responsible for the mechanical behavior variation with Ca/Si ratio. On the one hand, due to depolymerization role of Ca atoms, mean silicate chain length becomes shorter and Si–O_B–Si bonds are significantly substituted by the Si–O_{NB}–Ca bonds. Previous first principle study on the tobermorite structure demonstrates that Si–O bond energy is larger than that of Ca–O bond in the C–S–H mineral analogue [32]. Due to strength contribution from the long silicate chain, mechanical properties of C–S–H sample with low Ca/Si ratio are better than those with high concentration. However, in C–S–H structure, short silicate chains, such as dimmers or monomers, occupy predominate percentage, extremely weakening the tension strength and ductility. Murray's [17] study stated that the tensile strength of C–S–H can dramatically decrease to one-third of the original value, while the infinite silicate chains convert to dimmers completely. Analogously, in other calcium silicate composition system, such as silicate glass and geopolymer material, the silicate chain length also plays significant role in strength development. On the other hand, at high Ca/Si ratio, increasing number of water molecules penetrate in the cavities of the calcium silicate sheet, substituting the ionic-covalent bond with the unstable H-bond. The diffusion of interlayer water molecule results in the frequently breakage and formation of the H-bonds, which in great extent reduces the stability of the C–S–H gel.

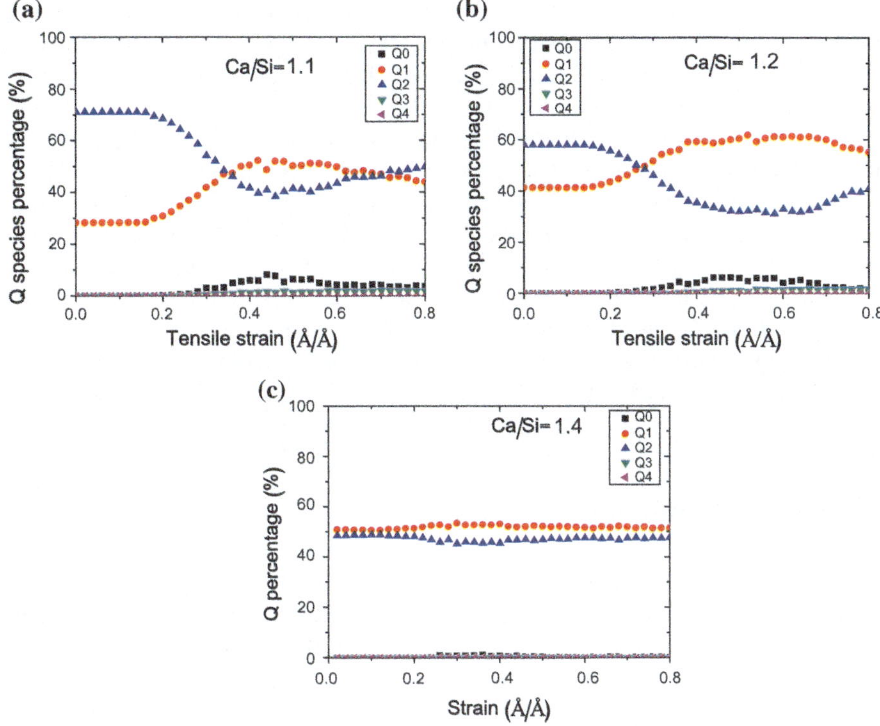

Fig. 4.23 Q species evolution of C–S–H gel with Ca/Si ratio **a** 1.1; **b** 1.2; **c** 1.4

Depolymerization reactions

To describe the silicate chain morphology changes, the Q_n percentage is calculated to describe the structural variation of the silicate skeleton during the tensile process, as shown in Fig. 4.23.

The long silicate chain in the C–S–H gel with low Ca/Si ratio varies greatly during the tension process. In the previous section, the silicate morphology evolution for the infinite chain in the tobermorite has been discussed. The silicate chain has similar evolution when Ca/Si ratio is quite low. As shown in Fig. 4.23a, the Q species evolution at Ca/Si = 1.1 can be clearly distinguished in three stages as the strain increases along the y-direction. Initially, the percentage of Q species maintains unchanged during the first 0.15 Å/Å strain period, in which stage, the silicate chain length is elongated and Si–O–Si angle is stretched open to carry the loading. Subsequently, when the strain exceeds 0.15 Å/Å, Q_2 begins to decrease, and Q_1 and Q_0 increases. Long silicate chains are broken into the short dimmers and monomers. Meanwhile, the Q_3 species grows at strains of 0.3 Å/Å, implying formation of the branched structures. Some silicate chains that were stretched broken can reconnect to form a new silicate skeleton. The newly connected branches can bridge the neighboring calcium silicate layers and construct the threedimensional structure. Different from the silicate chains in tobermorite, the ratio of Q_3 species are quite low, which weakens the plasticity for the molecular structure. Finally, the ratio of Q_2 slightly increases and Q_1 turns to decrease, as the strain reaches 0.46 Å/Å, from which the stress continues to reduce. It can be observed in Fig. 4.23b, c that as Ca/Si ratio increases from 1.1 to 1.4, the changed percentage of Q_2 species reduces from 33 to less than 5%, indicating weaker mechanical contribution from the defective silicate chains. However, when the Ca/Si ratio exceeds 1.4 or the mean silicate chain length reduces to 4, the Q species does not change, implying no depolymerization during tension along the y-direction for the C–S–H gel. Defective silicate chains, isolatedly distributing in the C–S–H gel, have not been drawn broken. Due to the missing bridging silicate tetrahedrons, the mechanical contribution from the Si–O bonds is significantly reduced. Hence, the tensile strength and stiffness of the C–S–H gel in the y-direction is weakened to a great extent.

4.7   Chapter Summary

Molecular simulation was utilized to construct the model of C–S–H gel. The validated MD model was applied to study the influence of Ca/Si ratio on the strength and mechanical properties of C–S–H gel. Several conclusions can be drawn from this study as flow:

(1)   The structural information of the simulated C–S–H gel model matches well with the data obtained by the X-ray diffraction and the Infrared spectroscopy and the mechanical properties calculated by reactive force field also is consistent with the values from the nanoindentation test.

(2)   The percentage of Q_0 species is controlled to less than 5%. The silicate skeleton contributes to a more reasonable local environment for the Ca atoms including the average Ca–O bond distance at 2.46 Å and the CN value 6.8, which matches well with the parameters found in the crystal chemistry of the general calcium silicate phases.

(3)   In the tensile loading process, the amorphous dry sample, due to the stretch of silicate branch and breakage of Ca_s–O_s bonds, demonstrates high stiffness, high strength, and plasticity in post-failure stage. Contrarily in a saturated state, much weaker Ca_w–O_s bonds and H-bonds play a major role in loading resistance, which leads to lower stiffness and strength, and makes the gel more brittle.

(4)   Different Young's modulus and tensile strength values obtained by uniaxial tension testing indicate heterogeneous mechanical performance. The high stiffness and cohesive force along the y-direction is mainly contributed by the silicate chains, while the weakest z-direction behavior is attributed to the frequently broken H-bonds network.

(5)   The reactive force field combines both the mechanical response and chemical response during the large tensile deformation process. On the one hand, silicate chains, acting in a skeletal role in the layered structure, depolymerize thereby

enhancing the loading resistance. On the other hand, water molecules, attacking the Si–O bond and Ca–O bonds, dissociate into hydroxyls, which are detrimental for the cohesive force development.

(6) With increasing Ca/Si ratio, tensile strength and Young's modulus of C–S–H samples gradually decrease for two aspects. Defective silicate chains at higher Ca/Si ratio weaken the loading resistance ability. In addition, increasing amount of interlayer water at high Ca/Si ratio further substitutes the Ca–O and Si–O with H-bonds network, resulting in the worse mechanical performance.

References

1. Cong, X., & Kirkpatrick, R. (1996). ^{29}Si MAS NMR study of the structure of calcium silicate hydrate. *Advanced Cement Based Material, 3*(3–4), 144–156.
2. Janika, J. A., Kurdowsk, W., Podsiadey, R., & Samset, J. (2001). Fraxtal structure of CSH and tobermorite phases. *Acta Physica Polonica, 100*, 529–537.
3. Allen, A. J., Thomas, J. J., & Jennings, H. M. (2007). Composition and density of nanoscale calcium silicate hydrate in cement. *Nature Material, 6*, 311–316.
4. Merlino, S., Bonnacorsi, E., & Armbruster, T. (2001). The real structure of tobermorite 11 Å: Normal and anomalous forms, OD character and polytypic modifications. *European Journal of Mineralogy, 13*(3), 577–590.
5. Hamid, S. A. (1981). The crystal structure of the 11 Å natural tobermorite $Ca_{2.25}[Si_3O_{7.5}(OH)_{1.5}]\cdot H_2O$. *Zeitschrift fur Kristallographie, 154*(3–4), 189–198.
6. Bonnacorsi, E., Merlino, S., & Taylor, H. (2004). The crystal structure of Jennite $Ca_9Si_6O_{18}(OH)_6\cdot 8H_2O$. *Cement and Concrete Research, 34*(9), 1481–1488.
7. Pellenq, R. J. M., Kushima, A., Shahsavari, R., Van Vliet, K. J., Buehler, M. J., & Yip, S. (2009). A realistic molecular model of cement hydrates. *PNAS, 106*(38), 16102–16107.
8. Shahsavari, R., Pellenq, R. J. M., & Ulm, F. J. (2011). Empirical force fields for complex hydrated calcio-silicate layered materials. *Physical Chemistry Chemical Physics, 13*(3), 1002–1011.
9. Youssef, M., Pellenq, R. J. M., & Yildiz, B. (2011). Glassy nature of water in an ultraconfining disordered material: The case of calcium silicate hydrate. *Journal of American Chemistry Society, 133*(8), 2499–2510.
10. Bonnaud, P. A., Ji, Q., Coasne, B., Pellenq, R. J.-M., & Van Vliet, K. J. (2012). Thermodynamics of water confined in porous calcium-silicate-hydrates. *Langmuir, 28*(31), 11422–11432.
11. Ji, Q., Pellenq, R. J. M., & Van Vliet, K. J. (2012). Comparison of computational water models for simulation of calcium silicate hydrate. *Computational Material Science, 53*(1), 234–240.
12. Qomi, M. J. A., Ulm, F. J., & Pellenq, R. J. M. (2012). Evidence on the dual nature of aluminum in the calcium-silicate-hydrates based on atomistic simulations. *Journal of the American Ceramic Society, 95*(3), 1128–1137.
13. Brough, A. R., Dobson, C. M., Richardson, I. G., & Groves, G. W. (1994). In situ solid-state NMR studies of Ca_3SiO_5: Hydration at room temperature and at elevated temperatures using ^{29}Si enrichment. *Journal of Materials Science, 29*(15), 3926–3940.
14. Richardson, I. G. (2013). The importance of proper crystal-chemical and geometrical reasoning demonstrated using layered single and double hydroxides. *Acta Crystallographica Section B, 69*, 150–162.
15. Thomas, J. J., Chen, J., Jennings, H. M., & Neumann, D. A. (2003). Ca–OH bonding in the C–S–H gel of tricalcium silicate and white Portland cement pastes measured by inelastic neutron scattering. *Chemistry of Materials, 15*(20).

16. Manzano, H., Moeini, S., Marinelli, F., van Duin, A. C. T., Ulm, F. J., & Pellenq, R. J. M. (2011). Confined water dissociation in microporous defective silicates: Mechanism, dipole distribution, and impact on substrate properties. *Journal of the American Chemistry Society, 134*(4), 2208–2215.
17. Murray, S. J., Subramani, V. J., Selvam, R. P., & Hall, K. D. (2010). Molecular dynamics to understand the mechanical behavior of cement paste. *Journal of the Transportation Research Board, 2142*(11), 75–82.
18. Dolado, J. S., Griebel, M., & Hamaekers, J. (2007). A molecular dynamic study of cementitious calcium silicate hydrate (C–S–H) gels. *Journal of American Ceramic Society, 90,* 3938–3942.
19. Puibasset, J., & Pellenq, R. J. M. (2008). Grand canonical Monte Carlo simulation study of water adsorption in silicalite at 300 K. *The Physical and Chemistry B, 112*(20), 6390–6397.
20. Gmira, A. (2003). *Etude texturale et thermodynamique d'hydrates modèles du ciment.* Orléans.
21. Janik, Y., Kurdowski, W., Podsiadly, R., & Samseth, J. (2001). Fractal structure of CSH and tobermorite phases. *Acta Physica Polonica Series A, 100*(4), 529–538.
22. Yu, P., Kirkpatrick, R. J., Poe, B., McMillan, P. F., & Cong, X. (1999). Structure of calcium silicate hydrate (C–S–H): Near-, mid-, and far-infrared spectroscopy. *Journal of the American Ceramic Society, 82*(3), 742–748.
23. Costantinide, G., & Ulm, F. (2006). The nanogranular nature of C–S–H. *Journal of Mechanics and Physics of Solids, 55*(1), 64–90.
24. Gelb, L. D., & Gubbins, K. E. (1998). Characterization of porous glass: Simulation models, adsorption isotherms, and the Brunauer–Emmett–Teller analysis method. *Langmuir, 14,* 2097–2111.
25. Cormack, A., & Du, J. (2001). Molecular dynamics simulations of soda-lime-silicate glasses. *Journal of Non-Crystalline Solids, 283–289.*
26. Mead, R. N., & Mountjoy, G. (2006). A molecular dynamics study of the atomic structure of $(CaO)_x(SiO_2)_{1-x}$ glasses. *Journal of Physical Chemistry, 110*(29), 273–278.
27. Grimley, D. I., Wright, A. C., & Sinclair, R. N. (1990). Neutron scattering from vitreous silica IV. Time-of-flight diffraction. *Journal of Non-Crystalline Solids, 119*(1), 49–64.
28. Mastelaro, V. R., Zanotto, E. D., Lequeux, N., & Cortes, R. J. (2000). Relationship between short-range order and ease of nucleation in $Na_2Ca_2Si_3O_9$, $CaSiO_3$ and $PbSiO_3$ glasses. *Journal of Non-Crystalline Solids, 262*(1–3), 191–199.
29. Pellenq, R. J. M., Lequeux, N., & Damme, H. V. (2008). Engineering the bonding scheme in C–S–H: The iono-covalent framework. *Cement and Concrete Research, 38*(2), 159–174.
30. Feuston, B. P., & Garofalini, S. H. (1990). Oligomerization in silica sols. *Journal of Physics and Chemistry, 94*(13), 5351–5356.
31. Shahsavari, R., Buechler, M. J., Pellenq, R. J. M., & Ulm, F. J. (2009). First-principles study of elastic constants and interlayer interactions of complex hydrated oxides: Case study of tobermorite and jennite. *Journal of American Ceramic Society, 92*(10), 2323–2330.
32. Zhu, T., Li, J., Lin, X., & Yip, S. (2007). Stress-dependent molecular pathways of silica–water reaction. *Journal of Mechanics and Physics of Solids, 53*(7), 1597–1623.
33. Cong, X., & Kirkpatrick, R. (1996). ^{29}Si and ^{17}O NMR investigation of the structure of some crystalline calcium silicate hydrate. *Advances in Cement Based Materials, 3,* 133–143.
34. Chen, J. J., Thomas, J. J., Taylor, H. F. W., & Jennings, H. M. (2004). Solubility and structure of calcium silicate hydrate. *Cement and Concrete Research, 34,* 1499–1519.
35. Powers, T. C., & Brownyard, L. (1946–1947). Studies of the physical properties of hardened Portland cement paste. *ACI Journal Proceedings, 43.*
36. Feldman, R. F., & Sereda, P. J. (1968). A model for hydrated Portland cement paste as deduced from sorption-length change and mechanical properties. *Matériaux et Construction, 1*(6), 509–520.
37. Kumar, R., Schmidt, J. R., & Skinner, J. L. (2007). Hydrogen bonding definitions and dynamics in liquid water. *The Journal of Chemical Physics, 126*(20), 204107.
38. Pelisser, F., Gleize, P. J. P., & Mikowski, A. (2012). Effect of the Ca/Si molar ratio on the micro/nanomechanical properties of synthetic C–S–H measured by nanoindentation. *The Journal of Physical Chemistry C, 116*(32), 17219–17227.

Chapter 5
Molecular Simulation of Water and Ions Migration in the Nanometer Channel of Calcium Silicate Phase

C–S–H gel has a multi-scale porous structure that contains capillary pores and gel pores. The water and ion migration in the C–S–H gel determines the strength, creep, shrinkage, chemical and physical reactivity of cementitious materials. In this chapter, the transport properties of water and ions in the gel pore are systematically studied.

5.1 Introduction

Ionic transport behavior in the nanometer-scale channel is ubiquitous and important for a broad range of natural and synthesis phenomenon: ions passing through macromolecular pores in cell membranes for various biological functions [1], ionic selectivity in the nanoporous material for the purification or desalination of seawater [2], detrimental ions penetration into reinforcement concrete for durability damage of the material [3], ions trapped in the nanoporous zeolite structures for the catalytic property modification, ions confined in carbon nanotube for hydrogen storage [4], etc. In the bulk solution, the ions can associate with neighboring water molecules by H-bonds or ionic bonds to construct the hydration shell. When ions and water transport into the nanometer-scale channel that has a size comparable with ions–water cluster, the molecular structure of the hydration shell is significantly distorted. As compared with bulk water, the ions and water ultra-confined in the nanometer pores demonstrate dramatically different hydration structure, H-bonds network, and diffusion coefficients. These structure and dynamics features are greatly dependent on the electronic and geometric restrictions, and hydrophilic and hydrophobic nature from the ultra-confined substrate. Even though increasing attention has been drawn by understanding the nanoconfined ions in many disciplines, the chemical and physical nature of ions transport in nanopore has not been comprehensively interpreted.

The nanometer-scale gel pores in the Calcium–Silicate–Hydrate (C–S–H) provide a good medium to study the molecular structures and dynamics of water and ions in the ultra-confined spacing. As a microporous and meso-porous material, the C–S–H gel is constructed by the assembly of ordered and disordered calcium silicate sheets,

© Science Press and Springer Nature Singapore Pte Ltd. 2020
D. Hou, *Molecular Simulation on Cement-Based Materials*,
https://doi.org/10.1007/978-981-13-8711-1_5

with water and ions exchange in the interlaminar space with size from nanometer to micrometer [5]. The transport of the ions and water in the nanometer channel of C–S–H gel determines the mechanical properties and durability of concrete material. The detrimental ions invasion, reacting with hydration products and inducing corrosion of reinforcement, weakens the cohesive strength of cement-based material and destroys the microstructure for the material. Experimental techniques can be utilized to study the water confined in the gel pore of C–S–H include nuclear magnetic resonance (NMR) relaxation measurements, solid-state ^{1}H NMR experiments [6] and quasi-elastic neutron scattering (QENS) [7]. The QENS [8] distinguished the water in the Portland cement and tri-calcium silicate into three states: liquid water, chemical bound water, and physically adsorbed water that are primarily associated with C–S–H in the small nanopores. The water confined in nanometer channel demonstrates different dynamic quantities such as residence time in the pores, diffusion coefficients which were characterized by the Neutron Scattering (NS) [9], Broadband Dielectric Spectroscopy (BDS) [10] and proton field-cycling relaxometry (PFCR) [11].

However, investigating the water structure and dynamics by experiments alone is challenged by some limitations such as the material purity and instrument accuracy at the relevant length and timescales. Computational methods can help to interpret the experimental results and play a complementary role in understanding the structural and dynamic properties at the molecular level. Molecular dynamics (MD), a force field based computational method, is able to give a more quantitative illustration on the structure, dynamics, and energy of solid–liquid interfaces.

Water and ions in charged or neutral silicate or clay nanopores mimicking the gel pores in the C–S–H have been investigated by computational methods for many years. Their efforts play a guiding role in exploring the complicated C–S–H interface. By using the MD method, Debenedetti, Rossy and their coworkers investigated the effect of surface polarity on the structural, dynamics, and mechanical properties of the water confined between silicate plates [12, 13]. Their work showed that the structure of the surface water molecules is significantly influenced by the hydrophilic or hydrophobic nature of the silicate surface. While the slow diffusion rate of water near the hydrophobic surface is attributed to the ice-like water layer template by the silicate crystal, the slow dynamics near the hydrophilic surface is caused by the strong-H-bond at the interface region. The hydrophobic to hydrophilic transition has also been simulated by Siboulet, Dufreche, and their coworkers [14, 15]. The effect of the transition can be characterized by the normal self-diffusion coefficients of the surface water molecules. Furthermore, Grand Canonical Monte Carlo (GCMC) molecular simulations are an effective method to analyze the thermodynamics of the water in nanoporous silica [16, 17]. The simulated isotherms describing adsorption and desorption of water molecules in the nanopore match well with the experimental observations. The analysis of the fluid structure in the mesoporous glass provides molecular insights into confinement, disorder, hydroxylated effects, and the morphologies influences on water adsorbed on the hydrophilic surface of porous silicate glass. On the other hand, the interface between the water and clay phase has been investigated by Rotenberg et al. [18, 19]. They first gave a microscopic description of the exchange of water and ions between clay interlayers and microporosity.

Even though the contributions made for water confined in silica or clay nanopores are significant, a crystal or amorphous silica surface cannot accurately characterize the structure of the complicated C–S–H surface in which the silicate chains develop in the dreierketten style and counterions are distributed. The cement–water interface simulation began to attract researchers' interest in recent years [20, 21].

Kalinichev et al. [21], whose group developed the ClayFF force field [22], introduced the force field model to a cement system and investigated multiphases in the cement-based material. In light of the mineral interface properties, MD simulation was performed to study the water confined between the brucite layers [23]. They gave a systematic picture that the water structure transforms from an ordered arrangement near the surface region to a disordered one in the bulk solution. The highly organized surface water molecules are influenced both by the surface geometry and the electrically charged surface. However, the simulation emphasized the brucite and water interface and did not show structural and diffusion rate details about the water on the surface of the tobermorite, jennite as well as the portlandite. Even though portlandite is the analogue for the brucite crystals, a small structural discrepancy exists in the crystal size and hydroxyls position. In addition to the water/cement interface study, they also performed MD simulation to study the structure and dynamic behavior of chloride at the interface between the cement hydrate and the ions solution [20]. Good descriptions, both in structure and dynamic properties of various cement phases, can confirm the suitability for the ClayFF force field in cement material simulation. However, the short timescale simulation (smaller than 0.5 ns) and the small number of ions (less than 10) cannot give reliable statistical results. The simulation system should be enlarged and elongated to guarantee statistical reliability.

Because of the structural similarity, tobermorite, as the important mineral analogues of cement hydrate, was used to investigate the transport behavior of the C–S–H gel at the molecular level, which are the preliminary studies. In this chapter, the adsorption model and transport model were constructed by the molecular dynamics method. While the former emphasizes the interaction between solution species and the C–S–H substrate, the latter focuses on the migration of ions and water in the nanometer channel.

5.2 Adsorption Model for Water and Ions Confined in C–S–H Gel Pore

The adsorption model aims to give insights on the molecular structure, dynamic motion, and interaction between the water and the calcium silicate surface. Density profiles, atomic trajectories, diffusion coefficients, and H-bond networks were employed to characterize the structural and dynamic behavior of water molecules in the nanopores. In addition, Cl^- and associated cations diffusion was also investigated in order to interpret the chemical and physical reactivity between the surface and the ions.

5.2.1 Computational Details

5.2.1.1 Model Construction

The simulation of the crystal structure is based on the 11 Å tobermorite structure [24]. In order to efficiently implement in LAMMPS, Murray [25] transformed from monoclinic coordinates to orthorhombic coordinates. The normalized orthorhombic structure is taken as the tobermorite substrate. Cleavage of the tobermorite structure in the [001] direction leads to non-bridging oxygen (O_{NB}). The surface Si–O$^-$ bonds which point perpendicularly to the substrate are transformed to Si–OH by adding hydrogen atoms. Therefore, the final tobermorite surface is terminated with Si–OH and Si–O$^-$. The silicate chains are assumed to be infinite with no defective part. The bridging oxygen atoms (O_B) and O_{NB} from the silicate chains become the predominant element in the channel. As shown in the cleaved structure in the *a*-direction, two columns of oxygen atoms lie at the bottom of the silicate channels, one of which is inside the O_B connecting the bridging silicon, and the pair silicon and the other is the O_{NB} from the bridging silicate tetrahedral. It can be observed in the *b* direction (Fig. 5.1) that the channel entries are located with Si–OH groups and in the following section analysis, the center of hydroxyl is defined as the surface line position.

After the completion of cleaving the supercell, two models with different interlayer solutions were constructed. For the first model, water molecules are added between the C–S–H substrates to construct a pore of width 4.5–6.5 nm. The selected pore width is in the range of gel pore sizes distributed from 0.5 to 10 nm, as mentioned by Mindess [26]. As proposed for most cases of water and ions simulation in nanopores [12, 20, 21, 23], the number of water molecules in the pore is assumed to satisfy the density of bulk aqueous solution under ambient conditions (~1 g/cm^3). The positions of the water molecules are initially randomly arranged in the pores. In the first case,

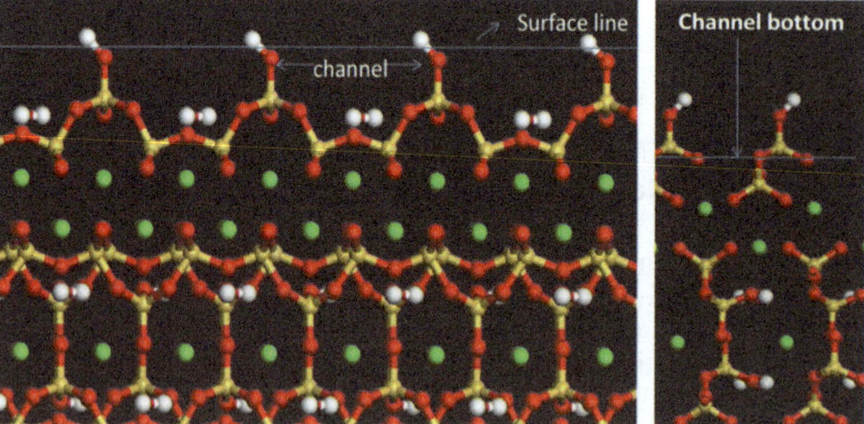

Fig. 5.1 The cleaved surface of the tobermorite structure

the interaction between water and tobermorite surface is investigated by analyzing the surface water density, orientation, H-bond distribution, and diffusion coefficient. In order to study the interaction between different ions, the $CaCl_2$ or NaCl solution is simulated by further adding Cl^- ions and Ca^{2+} (Na^+) into water molecules confined in the nanopore. Initially,all the ions are randomly located 20–25 Å away from either substrate surface. The influence of the ions is studied by analyzing the structural and dynamic properties of ions in the calcium silicate substrate/water interface. An example of the $CaCl_2$ solution diffusing in the nanopore is illustrated in Fig. 5.2.

5.2.1.2 Force Field and Molecular Dynamics Modeling

The ClayFF force field [22] is employed to describe the aqueous cation, water solution, and the cement-related phases. The time step was 0.001 ps and temperature was set to 300 K. During the simulation process, the thermodynamic properties, such as the pressure, temperature, and total energy, were carefully monitored. The simulation followed three steps: initially, the substrate structure was frozen and the whole solution system was relaxed for 100 ps; after water system equilibrium completion, the substrate and water were all relaxed over 300 ps that is sufficient time for the whole system to reach equilibrium. Finally, the 10,000 ps NVT running continued after the system equilibrium was reached. Every 0.1 ps, the configuration information was

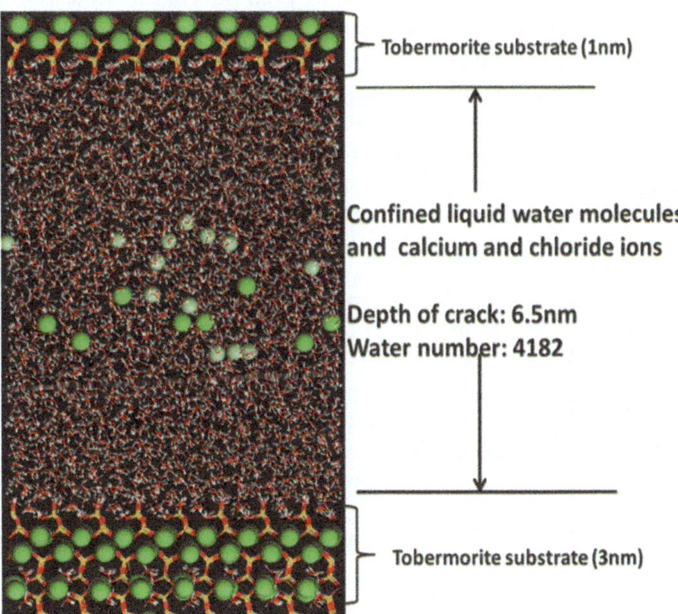

Fig. 5.2 Initial configuration for the water and ions confined between tobermorite substrates. The cyan balls are chloride ions

sampled, so as to analyze the equilibrium dynamic trajectory consisting of 100,000 atomic configurations. This large sample can guarantee statistically stability for data analysis.

5.2.1.3 Data Analysis

Atomic density, orientation, H-bonds, and diffusion coefficient profiles are important parameters for estimating the structure and dynamic properties of water molecules in nanopores. Atomic density profiles along the direction normal to the substrate surface were calculated by averaging 100,000 frames of the MD trajectories. Density profiles in both directions can reflect various energetic preference locations for the water or ions.

The orientation profiles, describing the water angle distribution, are defined by two special angles: φ_d and φ_{hh}. As illustrated in Fig. 5.3a, the former is the angle between the dipole vector (v_d) of a water molecule and the normal vector (v_n) for the surface, while the latter is the angle between v_{hh} (the vector from one H to the other in the same water molecule) and v_n. These instantaneous orientations of the water molecules at different layers show different tendencies that can be utilized to analyze the surface influence.

The H-bond network is important in characterizing the water structures and in describing the interaction between the substrate and the neighboring water molecules. H-bond network structures are constructed by both the contribution from the water molecules and the surface elements. Therefore, to describe local H-bond environments, it is necessary to calculate the average H-bond numbers at different distances away from the surface. As shown in Fig. 5.3b, the H-bond formation requires two conditions: two water neighbors with distance D_{HO} less than 2.45 Å and angle θ between the OH vector and OO vector less than 30°. If both conditions are satisfied, the one that provides H atoms to form the H-bond is called the donator, while the

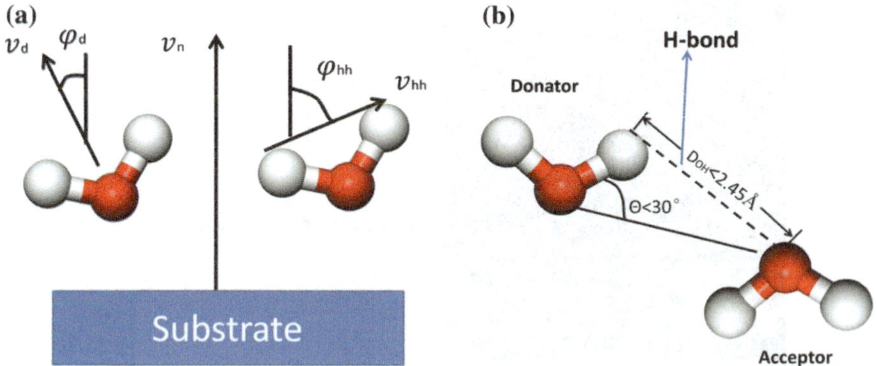

Fig. 5.3 **a** Water orientation: dipole angle and HH angle; **b** H-bond schematic illustration

oxygen container is defined as the acceptor. Similarly, the substrate element can also play the role of an H-bond donator or acceptor.

The diffusion coefficient, a parameter used to estimate the dynamic properties of water molecules, is achieved from the mean squared displacement (MSD), which has been introduced in Chap. 4. The positions of the water molecules are determined by the coordinates of the oxygen atoms from the equilibrium configuration. The MD trajectories were divided into 10 ps trajectories. Each recorded configuration was used as the origin of a new 10 ps trajectory. The interlayer water region was divided into several layers of width about 2 Å parallel to surface. The diffusion coefficient was calculated for each individual layer. For a typical 10 ns simulation, each layer coefficient was obtained from 10,000 10-ps configurations [27].

Ions and surface interaction can be described by the ion density profiles and ion diffusion coefficients. Ionic diffusion in the nanopores is complicated so that the analytical methods discussed above should be combined to explore the structure, dynamic, and energetic properties.

5.2.2 *Atomic Intensity and Orientation Files for Water Confined in Gel Pore*

For pore solutions containing water molecules and Ca ions, the structure and dynamic properties of the water in the gel pores were investigated. Density profiles normal to the substrate surface were used to give first insights of the water distributions. The center of the surface hydroxyl bond is defined as the entry for the silicate channel. For a simulation system with pure water (see Fig. 5.4a, b), the oxygen atoms profiles in water (O_w) have six peaks at 13.4, 15, 17.5, 19.5, 21.5, and 23.7 Å and oscillation becomes smaller up to 12 Å from the channel entry. Profiles of the hydrogen atoms in water (H_w) show four peaks at 14.4, 16.6, 18.6, and 20.7 Å, and the fluctuation gradually decreases up to 10 Å. Similar with tobermorite, the first peak of O_w is located at a distance about 1 Å smaller than that of H_w, implying that the tobermorite channel energetically prefers to adsorb the O_w atoms. The O_w and H_w density oscillation indicates strong layering of the water molecules in the vicinity of a solid surface. The spatial ordering of the water molecules confined in the nanopore has also been observed in the interface between water and other materials such as silica [16], brucite [23], and clay [18]. For example, increasing the local water density in the vicinity of both hydrophilic and hydrophobic surfaces has been reported in the previous simulation [12]. The ice-like water layer templated by the hydrophobic surface and the strong-H-bond connections in the hydrophilic surface result in the local density increasing. Only the first two water layers are strongly influenced by the silica or brucite surface, resulting in the two pronounced intensity peaks. However, in the tobermorite surface, four strong water layers indicate more complicated silicate geometry. Due to the large gap between the silicate chains, water molecules

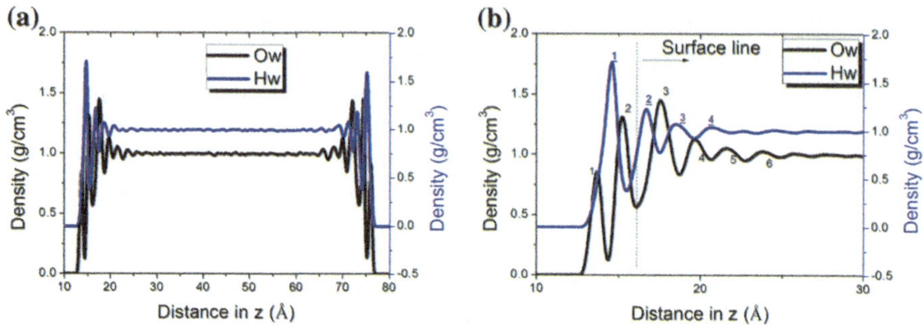

Fig. 5.4 **a** Atomic density profiles in the z direction ($z = 16$ Å is the surface line); O_w and O_h refer to the oxygen atom and hydrogen atoms in water; **b** is enlarged for better observation

can penetrate into the silicate channel and the channel water molecules contribute to another two obvious water layers.

The first and second peaks in the O_w density profile categorize channel water molecules into two types: type 1 and type 2. The two density distribution peaks are clearly distinguished, not overlapping as in the density profiles in jennite interfaces [28], implying a local constraint discrepancy between the two types of water molecules. The wide shoulder between the second and third layer is the location of the surface line and also indicates the boundary between the surface water and the channel water molecules. Compared with the jennite–water interface, the tobermorite interface has a high concentration of Si–OH and can lead to more water adsorption so that the surface water density, reflected by the third peak, is larger than that in jennite (1.2 g/cm^3). At a distance of 2 Å from the surface, the water density gradually decreases from 1.5 to 1 g/cm^3 as in the bulk water solution. The smaller variation continues up to 12 Å from the surface.

The water molecule dipole moment magnitude [29] can be used to characterize the hydrophilic or hydrophobic nature of tobermorite surfaces. While the downshift of the dipole moment distribution corresponds to the hydrophobic interaction of tobermorite surfaces, the upshift is hydrophilic. The dipole moment distribution of the first three layers in the O_w density distribution is plotted in Fig. 5.5. As compared with the average dipole moment of bulk water molecules (2.44 D), the value for the type 1 and type 2 channel water molecules are slight decreased and increased, respectively, while the value of surface water molecule remains unchanged. The slight change of dipole moment implies that the tobermorite surface has no strong hydrophilic or hydrophobic trend. As proposed in previous research, the bridging oxygen (O_B) atoms in the siloxane bond show a hydrophobic nature [30], while the non-bridging oxygen atoms [14] and hydroxylated silicate surface [16] exhibit hydrophilic behavior. Since O_{NB} and O_B atoms coexist and their percentage is similar near the tobermorite surface, neither hydrophobic nor hydrophilic preference is shown in the surface. The silicate hydroxyls are attributed to H-bonds linkage between the tobermorite and the water. In a real C–S–H structure, the surface has defective rather than infinite long chains as

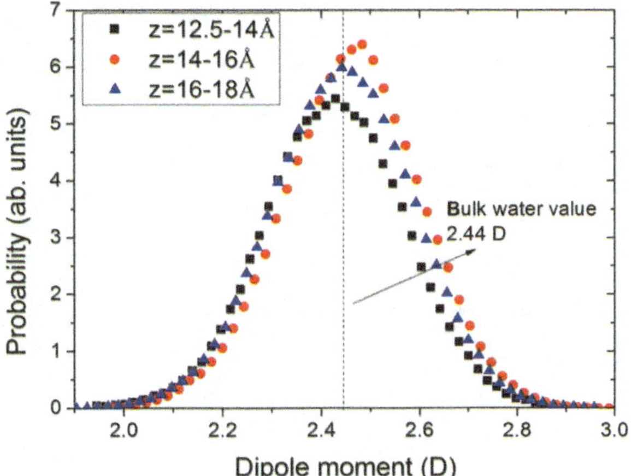

Fig. 5.5 Dipole moment magnitude distribution for the first three layers in water density distribution

in the tobermorite structure, and the dimmer structure has a predominant distribution. More O_{NB} atoms, which contribute to hydrophilic nature, result in more H-bonds connection with neighboring water molecules.

Surface and channel water organization can be well described by the atomic density contours projected in the surface plane. The triangular trajectories of Si–O tetrahedron are used as the reference positions for analysis. The trajectories of the water molecules of the first three peaks in the O_w intensity are recorded in Fig. 5.6. The triangular contours with silicon in the center are the projections of the bridging silicate tetrahedrons, which are taken as the reference positions for the water trajectories. As shown in Fig. 5.6a, the channel water molecules have well-ordered structures and the trajectories show that they are distributed in a line along the silicate chain extension (*b* axis). As shown in Fig. 5.6a, since few trajectories overlap two neighboring water molecules in different silicate channels, it is believed that the mobility of type 1 is relatively slow and no position exchange occurs. In the upper layer (Fig. 5.6b), type 2 water molecules are concentrated near the O_{NB} atoms of the silicate tetrahedron and the Ca atoms, indicating the strong electronic interaction between water molecules and Ca, O_{NB} atoms. At the distance away from the Ca and the silicate tetrahedron, the contours become less ordered and exchanges between the water molecules occur frequently. The third density map (Fig. 5.6c) exhibits the arrangement of the surface water. The less well-ordered surface water layer is also a transition zone, above which the ordered organization is difficult to be observed. The atomic trajectories from the bottom to the top layers apparently show that water molecules transform from orderly organized channel water to disordered bulk water.

The dipole and HH orientation of the type 1 channel water molecules, contributing to the first peak of the oxygen profiles, are shown in Figs. 5.7a and 5.8a. The dipolar angle distribution of type 1 water molecules is characterized by an intense peak at around 30° and a smaller peak at around 85°. The angle 30° indicates that type 1 water molecules have both OH-bonds pointing toward the channel entry. This can also be

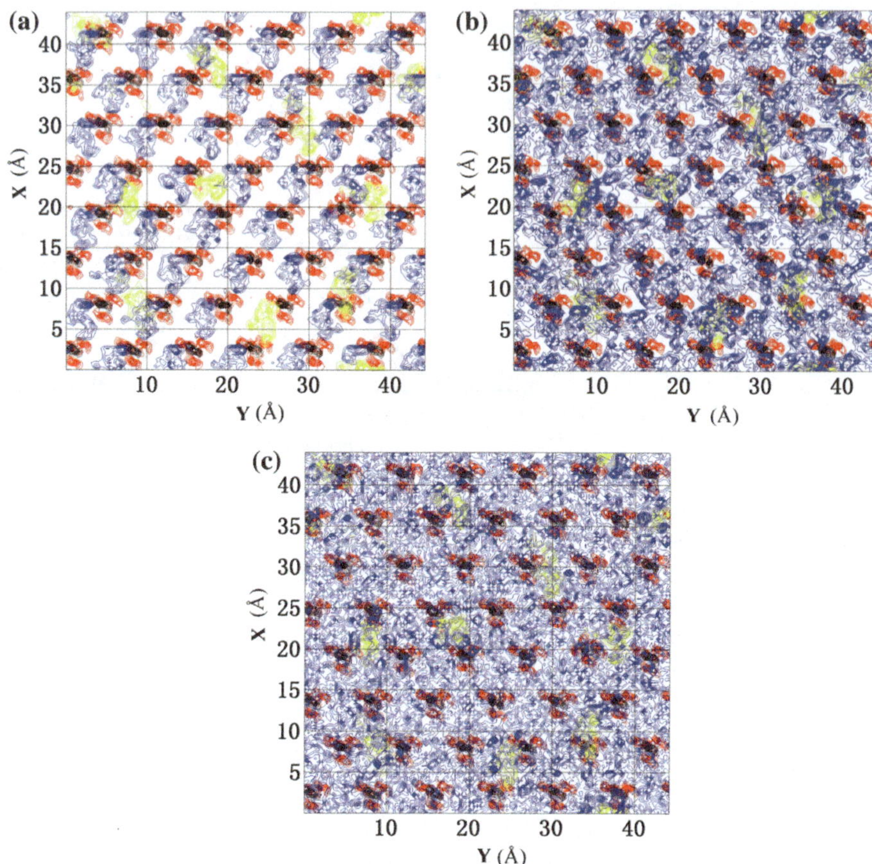

Fig. 5.6 Water molecules trajectories of the tobermorite and the water interface: **a** type 1 water molecules; **b** type 2 water molecules; **c** surface water molecules

confirmed by the fact that the first peak position in the hydrogen density profiles is 1 Å larger than that in the oxygen density profiles. The angle 85° corresponds to the dipolar vector of water molecules parallel to the tobermorite substrate. It can be observed from Fig. 5.7a that the φ_{hh} angle is distributed with a sharp peak at 90° and symmetric shoulders at 67° and 113°.

As shown in Fig. 5.8b, the water molecules, with z distances from 14 to 15 Å, have three peaks located at 30°, 97°, and 134°. While the first two peaks with smaller intensity resemble the angular distribution of type 1 water molecules, the angle 134° exhibits a typical feature of type 2 water molecules, which have dipole vectors pointing toward the tobermorite substrate. In the upper layer, as illustrated in Fig. 5.7c, the sharp and intense peak at 167° indicate that the type 2 water molecules are predominantly oriented with the dipolar axis pointing to the solid. Additionally, as shown in Fig. 5.8c, the φ_{hh} with a sharp peak at around 90° indicates that both H_w

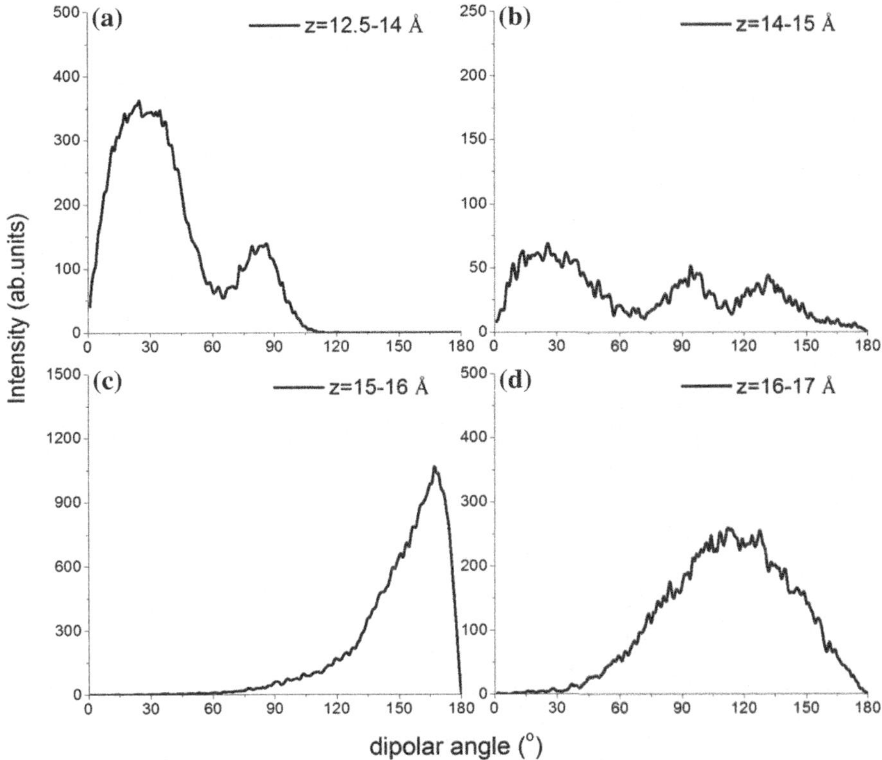

Fig. 5.7 φ_d distribution for the channel and surface water molecules **a** type 1 water molecules; **b, c** type 2 water molecules; **d** surface water molecules

atoms in the water molecules are closer to the solid surface than O_w atoms. In the surface water layer, shown in Figs. 5.7d and 5.8d, the broad peak of φ_d at 120° and the wide angle distribution of φ_{hh} implies more random orientation of the surface water molecules.

5.2.3 H-Bond Network and Coordinated Atoms

The channel water and surface water properties discussed above are determined by the interactions between the water molecules and the silicate channel. The tobermorite and water interface are connected by the H-bond network. There are three types of oxygen atoms near the interface, O_B from Si–O–Si, O_{NB} from Si–OH, and O_h from Si–OH, which can develop H-bond networks with neighboring water molecules. As shown in Fig. 5.9, with the progressively increasing distance from the channel bottom, the H-bond number increases monotonically to 3.89 at 1 Å below the channel entry,

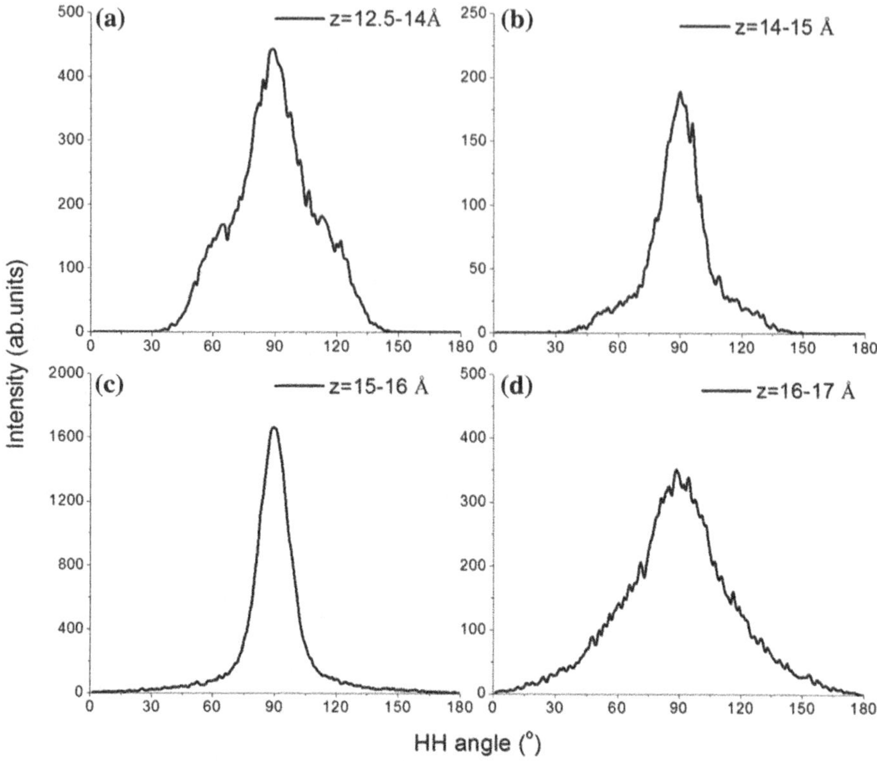

Fig. 5.8 φ_{hh} distribution for the channel and surface water molecules **a** type 1 water molecules; **b, c** type 2 water molecules; **d** surface water molecules

and decreases to 3.5 (bulk water value) around 10 Å from the surface, with a small fluctuation at 2 Å from the surface. For a type 1 water molecule, on average, the H-bond number is 2:0.95 and 0.2 H-bonds are donated to and accepted from neighboring water molecules, and 0.85 H-bonds are donated to O atoms in the silicate chain.

As shown in Fig. 5.10a, b, because the O_w atom is one of the nearest neighbors of the calcium layer, the OH vector points upward. Hence, it explains why $\varphi_d = \sim30°$ for type 1 water molecules. Meanwhile, due to the upward orientation, the major percentage of type 1 water molecules can only donate H-bonds to the higher layer water molecules and a small ratio of molecules accept H-bonds from neighboring water molecules. On the other hand, on average, type 2 water molecules accept 1.86 H-bonds from neighboring water molecules and donate 1.9 H-bonds to O_B or O_{NB} atoms in the silicate chains, as shown in Fig. 5.10c. Additionally, on average, type 2 water molecules can donate only 0.1 H-bonds to type 1 water molecules. As illustrated in Fig. 5.10c, the attraction from oxygen atoms located inside in silicate chains is the principal reason that leads to OH-bonds of type 2 water molecule point

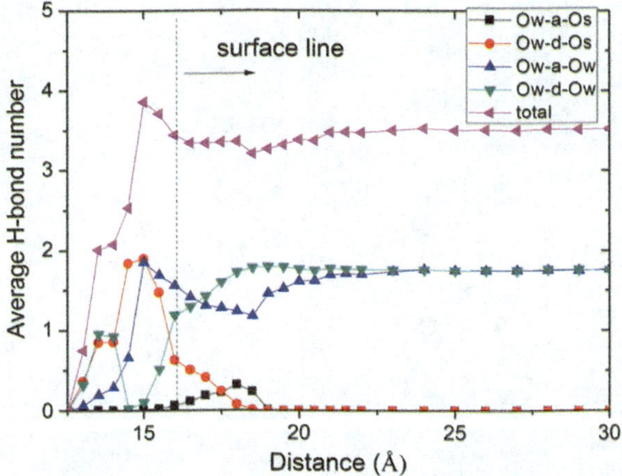

Fig. 5.9 H-bond number distribution and decomposed H-bond with contribution from water and surface element. O_w–d–O_w: H-bonds formed by water molecules that donate H-bonds to neighboring water molecules; O_w–a–O_w: H-bonds formed by water molecules that accept H-bonds from neighboring water molecules; O_w–d–O_s: H-bonds formed by water molecules that donate H-bonds to O_B, O_h or O_{NB} in the silicate chains. O_w–a–O_s: H-bonds formed by water molecules that accept H-bonds from O_B, O_h, or O_{NB} in the silicate chains

downwards. While type 2 water molecules diffuse between two silicate chains, both O_B from one silicate and O_{NB} from the other restrain the water orientation.

Nevertheless, as shown in Fig. 5.10d, the water molecules at the surface have less interaction with the silicate chains than that in the channel water. The ratio of H-bonds connected with O atoms in the silicate chains to those with neighboring water is larger than 1:5. The weak constraint from the surface results in a random distribution of surface water, as shown in the oxygen contours as well as the orientation profiles. Up to 10 Å from the surface, water molecules both donate and accept 1.75 atoms to their neighboring molecules (approximating to the value for bulk water).

The coordination oxygen atoms also influence the local environment for interface water molecules. Some coordinated oxygen atoms, not forming H-bonds with water molecules, contribute to defining the water molecule structures. As shown in Fig. 5.11, on average, the total coordination number per water molecule decreases from 6.5 to 5 at 14 Å and increases and reaches a peak value of 5.7 at around 15 Å, with a small fluctuation at the surface. It then decreases to 4.3 (a value similar to bulk water one) with 4 Å away from the surface. Figure 5.12 shows that besides the elements forming the H-bonds, the coordinated atoms also include 3–4 additional O_{NB} atoms from the silicate chains. These O_{NB} atoms and the O_w atoms in the type 1 water molecules are the nearest neighbors of the up layer calcium atoms and it is the reason that type 1 water molecule has similar mobile trajectories as the O_{NB} atoms in the silicate tetrahedron. As shown in Fig. 5.12b, for type 2 water, the coordination number is about 1.2 larger than the H-bond number. The extra coordinated atoms are mainly contributed by the nearest O_h atoms.

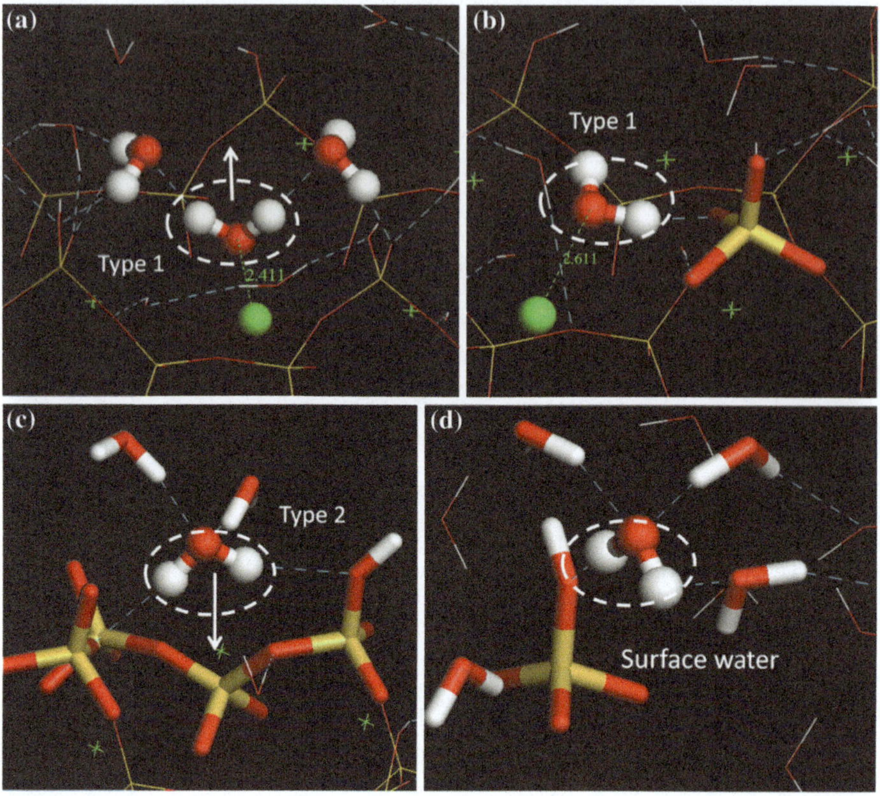

Fig. 5.10 **a**, **b** Local structure of type 1 water molecules; **c** local structure of type 2 water molecules; **d** local structure of surface water molecules. The dotted line represents the H-bonds

Fig. 5.11 Coordination number evolution with distance from the channel bottom. Cutoff distance for O–O coordination is 3.3 Å

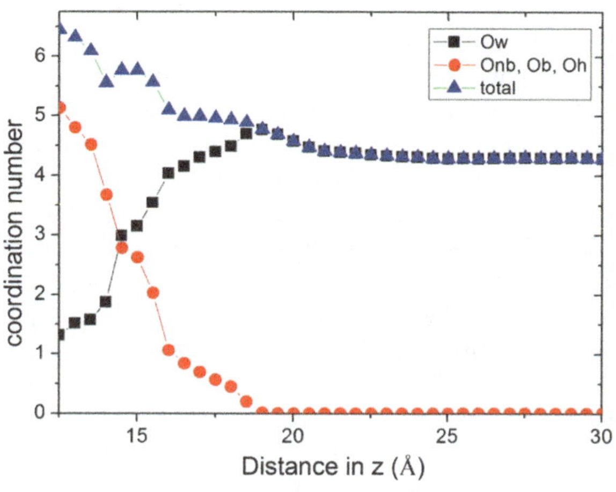

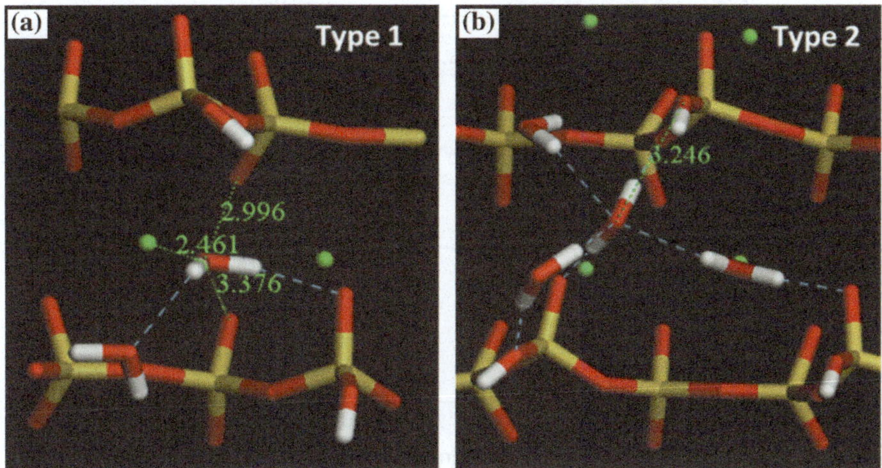

Fig. 5.12 Snapshots of the nearest neighbors of **a** type 1 and **b** type 2 water molecules

5.2.4 Diffusion Coefficient

The previous discussion showed that highly ordered channel water and surface water molecules are influenced by the tobermorite substrate. The diffusion coefficient, calculated from MSD over 10 ns configurations, reflects the dynamic properties of different water species. As shown in Fig. 5.13a, from the bottom of the silicate channel to the middle position of the nanopores, the crack is divided into four categories: the channel water layer (mainly type 1 and type 2 water), the surface water layer (within 10 Å from surface), the water layer 10–20 Å away from the surface, the water layer located in the center of the pore. The diffusion coefficients for the type 1 and type 2 water molecules are 0.15 and 1.1×10^{-9} m^2/s, which is around 1/26–1/4 the value of D_{H_2O}. The simulated diffusion coefficients are of the magnitude of the diffusion coefficient of water between the C–S–H particles tested by a proton field-cycling relaxometry approach (PFCR) ($D = 1/60\ D_{H_2O}$) [31] and time-resolved incoherent elastic neutron scattering (QENS) ($D = 1/6\ D_{H_2O}$) [32]. Because the PFCR and QENS are developed to detect the dynamic properties of water molecules adsorbed on the nanopore surface, MD simulation results are comparable to those obtained from experiments. This means that both the channel water and surface water exchange positions with neighboring molecules at a very slow rate. Especially for type 1 channel water, it is deeply embedded in the channel and only oscillates around the equilibrium position rather than diffusing randomly and being transported into the bulk solution. The extremely low diffusion rate also implies the glassy nature of the channel water molecules, which is a typical feature for structural water molecules ultra-confined in a layered structure [29].

Due to the quasi-two-dimensional geometry of the nanopores in the tobermorite, three types of diffusion coefficient are calculated: the three-dimensional diffusion

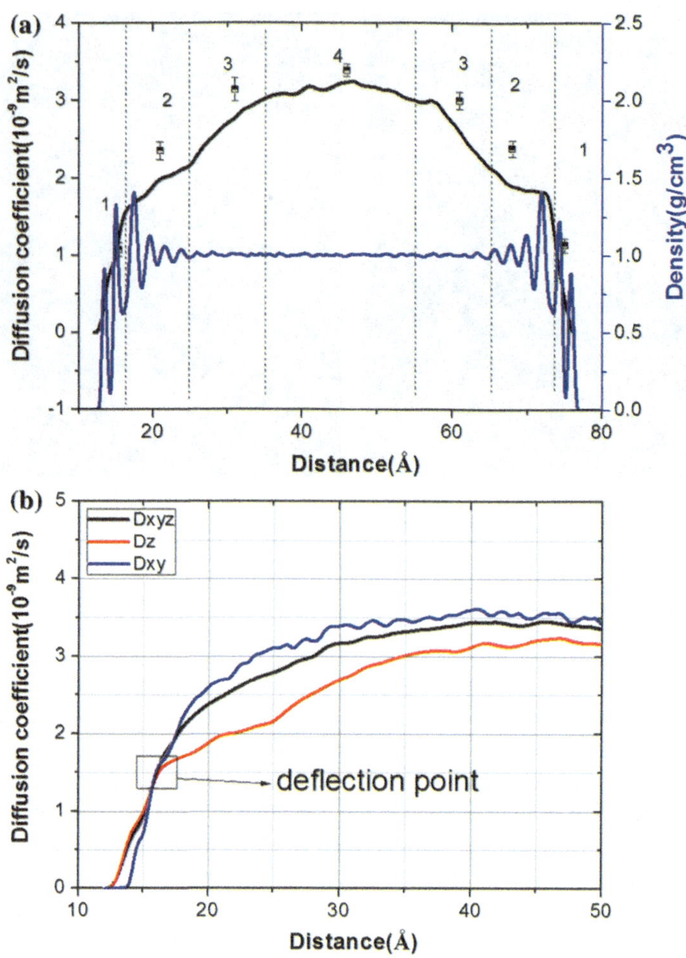

Fig. 5.13   **a** Diffusion coefficient changes with distance from the tobermorite surface; **b** D_z, D_{xy}, and D_z evolution with the distance from tobermorite surface

coefficient D_{xyz}, the two-dimensional coefficient parallel to the tobermorite substrate, D_{xy} and the one-dimensional coefficient normal to the tobermorite substrate, D_z. The pore space can be further decomposed into thinner layers normal to the z-direction so that the coefficient variation with the progressively increasing distance can be observed. The layer width is 1 Å and the maximum standard deviation of the diffusion coefficient is about $\pm 0.4 \times 10^{-9}$ m²/s. Longtime simulation can reduce the statistical uncertainties. As shown in Fig. 5.13b, the evolution curve D_z has a deflection point at the location of the silicate channel, same as the case of jennite. In the silicate channel, the molecular displacement of the water is almost isotropic in all directions. From the tobermorite surface, the discrepancy between D_{xy} and D_z increases and reaches a maximum around 10 Å from the surface. When the water molecules are located in the central position of the nanopores, the difference between D_{xy} and D_z becomes smaller. The variation of the diffusion coefficient parallel and normal to the tobermorite surface is attributed to the quasi-two-dimensional geometry for the nanopores. Compared with the constraint normal to the substrate, the connectivity in the XY plane contributes to the large diffusion motion of the water molecules.

5.2.5 Interaction Between Ions and Tobermorite Substrate

At the nanoscale, since the surface of C–S–H gel is highly negative, the cations diffusing between the nanopore form a high concentration solution. Figure 5.14a shows the intensity evolution of ions and water in the pore solution. According to the ion density profiles, less than 5% Cl^- diffuse within 4 Å from the surface. The relatively low density near the surface indicates that the tobermorite surface strongly repulses the chloride ions. These results are also consistent with previous computational results [20, 33]. In the real C–S–H structure, the concentration near the surface should be much smaller because the O_{NB} atoms from the defective silicate chains result in a more negative surface and the electronic repulsion force becomes stronger. On the other hand, a high-density peak of Ca^{2+} is exhibited between the channel water and the surface water. More than 66.5% of the total Ca^{2+} ions occupy a region within 4 Å of the surface. The predominant distribution of Ca^{2+} in this region implies that the tobermorite surface provides strong adsorption for the cations and explains that the positive surface potential of the early cement hydration product, which is attributed to adsorption of Ca^{2+} on the hydrate surface [34].

To further analyze the interface structure of the ions, the trajectories of the Ca^{2+} and Cl^- ions within 4 Å from the surface are recorded in the contour maps in Fig. 5.14b, c. The Ca^{2+} map demonstrates the high-intensity trajectories and the concentrated distribution above the silicate tetrahedrons, implying a relative stable bonding between the Ca^{2+} and O atoms in the silicate chains. However, the loose intensity for the Cl^- ions trajectories also confirms the repulsive role of tobermorite surfaces to anions. As shown in Fig. 5.15a, b, surface O_h atoms in the silicate chain play an important role in attracting the Ca^{2+} ions. Due to the high positive valence of Ca^{2+}, the strong electronic attraction force can result in a stably bound $Ca–O_h$. On the other hand, the channel and surface water molecules with O_w atoms exposed upwards are more likely to form $Ca–O_w$ connections. Furthermore, the diffusion coefficient of the surface calcium ions is less than 0.023×10^{-9} m^2/s, which is even smaller than the channel water molecules. It is suggested the $Ca–O_h$ connection is more stable than the H-bond connection $O_w–O_h$. However, because of no strong restriction from the surface, the chloride ions diffuse in a rapid mode with D equal to 1.73×10^{-9} m^2/s. It can be observed in Fig. 5.15c that Ca^{2+} and Cl^- clusters exist near the surface. These clusters, to some extent, increase the sorption of the Cl^- ions. For short time simulation (less than 100 ps), the Ca–Cl connection is quite stable and the cluster is located near the silicate channel entry. Nevertheless, for long timescale simula-

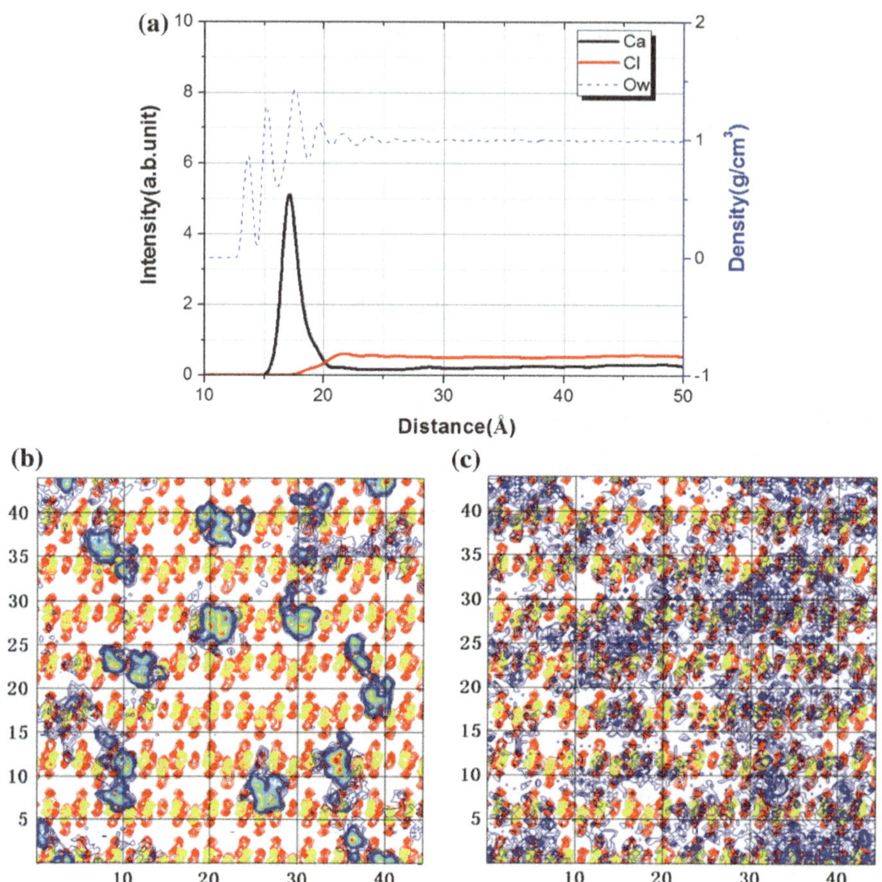

Fig. 5.14 **a** Snapshots for the Ca^{2+} and Cl^- ions that diffuse in the nanopores confined by the tobermorite surface; **b** local structure of Ca^{2+} adsorbed by the tobermorite surface; **c** Ca–Cl cluster formation near the surface

tion (larger than 2 ns), most of the surface calcium atoms diffuse from the silicate channel and move away from their original positions. It can also be observed that some calcium ions diffuse into the solution and are coordinated by O_w atoms rather than O_h atoms in the silicate chains. This calcium ions desorption from the surface has been mentioned in previous simulations [27]. Because the previous simulation utilized short simulation time and a stable original tobermorite structure, this desorption phenomenon is difficult to be observed. The coordinated Cl^- ions accelerate the desorption process.

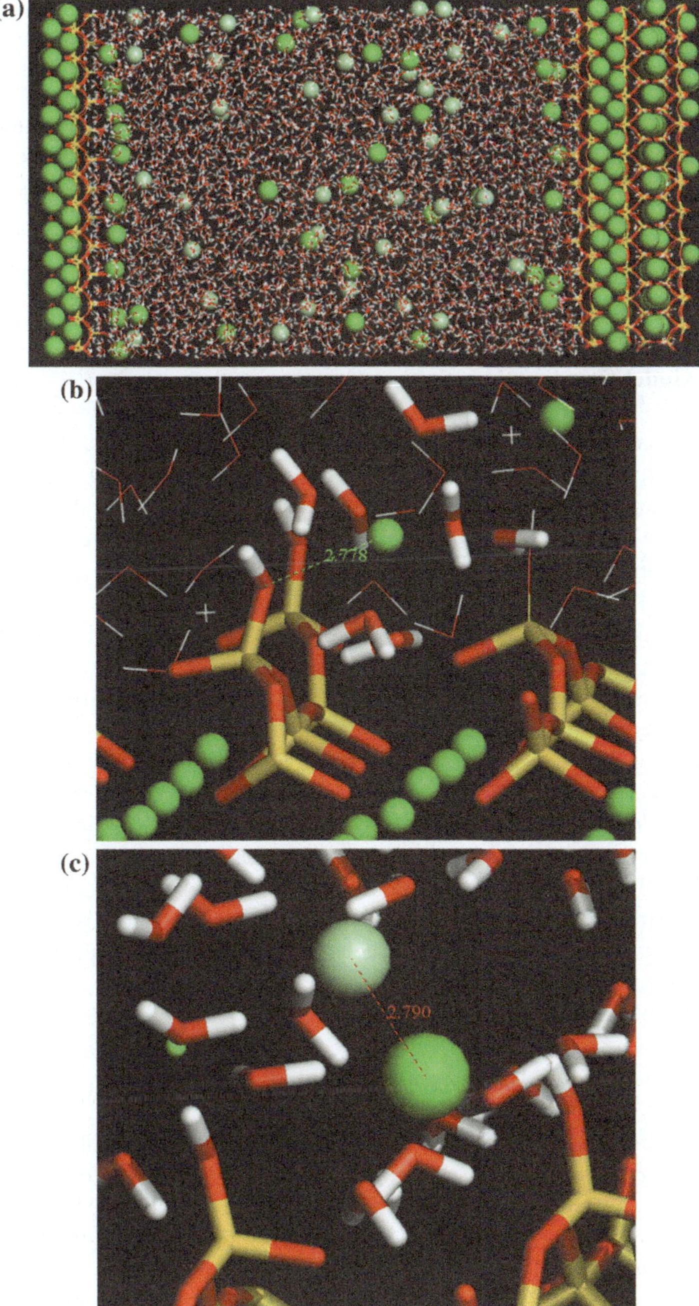

Fig. 5.15 a Snapshots for the Ca^{2+} and Cl^- ions that diffuse in the nanopores confined by the tobermorite surface; **b** local structure of Ca^{2+} adsorbed by the tobermorite surface; **c** Ca–Cl cluster formation near the surface

## 5.3	Capillary Transport Model for Ions and Water in the Gel Pore

The transport model investigates on the migration of ions and water in the nanometer pores. Different transport mechanisms of water and ions in the gel pore are proposed in this section.

### *5.3.1	Computational Details*

#### 5.3.1.1	Model Construction

The capillary adsorption model includes two parts: the sodium chloride solution and calcium silicate gel pore. It can be observed in Fig. 5.16 that 25,620 water molecules and 420 Na^+ and Cl^- ions, were randomly arranged in the bottom domain with the size of 11.16 Å × 215 Å × 318 Å, to model 1.2 mol/L the sodium chloride solution with density around 1.04 g/cm^3. The other part is the C–S–H gel pore constructed by two parallel calcium silicate sheets with the thickness around 1 nm, which were arranged perpendicular to sodium chloride solution. The entrance of the gel pore was in contact with NaCl solution and the gel pore grew along y-direction for more than 200 Å. To consider the pore size influence, the distance between two neighboring C–S–H sheets was set as 35, 45, and 60 Å. The C–S–H surfaces was obtained by cleaving one of the C–S–H mineral analogue, the tobermorite 11 Å crystal along [001] direction as mentioned in Ref. [35]. Non-bridging oxygen atoms in the silicate tetrahedron were transformed to silicate hydroxyl and counter ions, interface calcium ions, remained binding with silicate chains, which can maintain the charge balance. Furthermore, the periodic boundary condition (PBC) was set in x- and z-directions for the simulation box. In nonperiodic y-direction, two invisible walls were fixed on the top and bottom surface, respectively, which avoid the escaping of the water molecules along y-direction.

#### 5.3.1.2	Force Field and Molecular Dynamic Procedure

The ClayFF force field [22] was utilized to model the atomic potentials of the atoms in the solution and calcium silicate skeleton. LAMMPS [36] was utilized to perform the molecular dynamic simulation in the following procedures: first, the C–S–H gel pore was "frozen" by rigid body algorithm but the atoms in the sodium chloride solution were relaxed under NPT ensemble at 300 K and at 1 atm. Meanwhile, an invisible wall was arranged at the interface between solution and C–S–H gel pore to avoid the water molecules and ions diffusing from the bulk solution. The solutions were relaxed for 500, 600, 800, and 1000 ps to obtain four different initial configurations for subsequent capillary adsorption simulation. Second, when the solution system

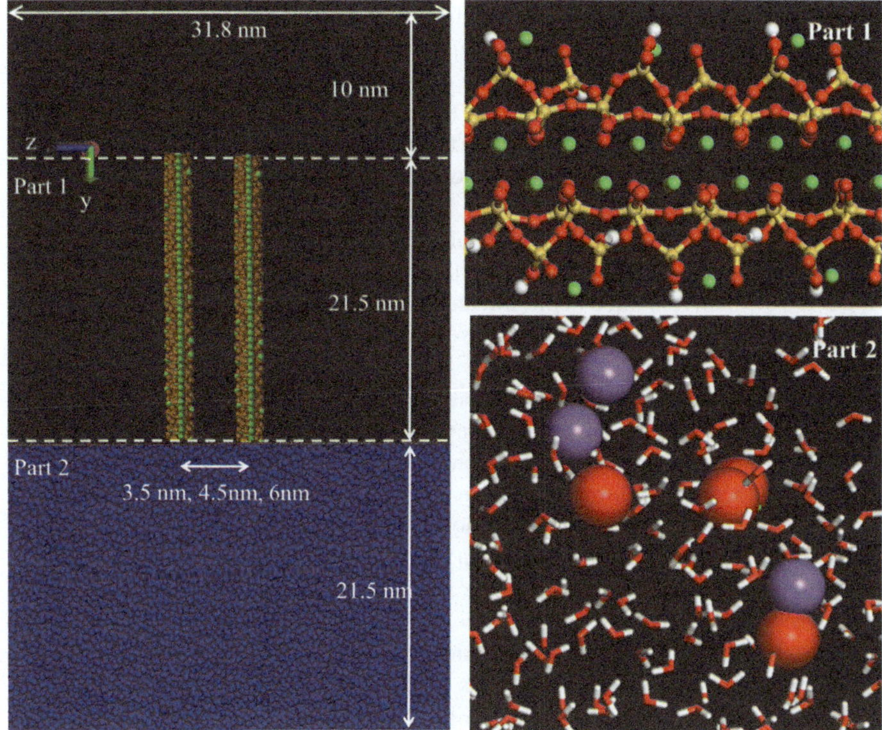

Fig. 5.16 Molecular model of solution transport in the slit tobermorite 3.5 nm channel. In the part 1, the green ball represents calcium ions. The yellow-red sticks are silicate chains, the red-white sticks are hydroxyl groups. In the part 2, the purple and red balls are the sodium and chloride ions. The white-red sticks are water molecules

reached an equilibrium state, the rigid C–S–H gel substrates were relaxed by 100 ps NVT runs so that the molecular structures were optimized for silicate hydroxyl and surface calcium ions. Finally, the flat invisible wall at the interface was removed and water and ions can transport through the gel pore freely for another 2000 ps during which the trajectories of atoms are recorded for statistical analysis. During this stage, the NVT ensemble was employed. The time step was set as 1 fs. The configuration information was recorded for data analysis every 100 fs. According to the initial solution configurations, the capillary adsorption process was repeated for four times. The data analysis based on four cases can give more statistical stable result.

5.3.2 Capillary Adsorption of NaCl Solution

It can be observed in Fig. 5.17a progressive imbibition for the water molecules into the nanometer channel of C–S–H substrates during 2000 Ps, taking 4.5 nm gel pore for example. It is clearly observed the advancing meniscus of the water in the nanopore, which is a typical feature for capillary adsorption. The downward curve of the advancing frontier with contact angle smaller than 90° is due to the hydrophilic nature of C–S–H surface. On the other hand, Fig. 5.17b records the sodium and chloride ions invasion into the channel from the bulk solution to the nanometer channel at 100, 500, and 2000 ps. As compared with Fig. 5.17a, the sodium and chloride ions migrate more slowly than water molecules. After the transport for 2000 Ps, the advancing frontier for different species in solution rank in the following order: water > Cl$^-$ > Na$^+$. Part of the surface calcium atoms was substituted by the sodium ions, which remained in the silicate channel. The dissociated surface calcium atoms can also diffuse into the gel pore solution and form ionic pairs with chloride ions.

The temporal penetration depth for water molecules has been calculated by monitoring the trajectories of the water frontiers. As exhibited in Fig. 5.17, solution imbibition frontiers are defined as the center of the capillary meniscus. Figure 5.18 demonstrates the water imbibition depth as a function of time for the pore with a diameter 4.5 nm. Initially, the temporal evolution of water penetration depth follows linear relation with time, implying the constant velocity regime from 0 to 250 ps for the capillary adsorption. At a later stage, the capillary meniscus advancing progressively departs from the constant velocity regime and follows a visco-inertia regime that is defined by the classic Lucas–Washburn equation (LW) [37] of capillary adsorption of water. The classic LW equation describes the capillary filling process to be a square root dependent of time. The water imbibition can be considered as an equilibrium state between the viscous force and capillary force [37].

$$y(t) = \sqrt{\frac{r\sigma \cos \theta}{2\mu}} \tag{5.1}$$

In Eq. (5.1), y represents the penetration depth, r is the radius for the nanopore (22.5 Å), σ is the surface tension of water molecules (for the SPC water model is 0.0547 N m^{-1}) and μ is the viscosity for water (for the SPC water model is 0.54–0.58 mPa s) and θ is the contact angle between water and tobermorite surface (45°). The penetration depth based on equation during 2000 ps was calculated as the function of time in Fig. 5.18 and the positions of the simulated capillary advancing front were also listed for comparison. It can be observed in Fig. 5.18 that the capillary front follows a parabolic visco-inertia regime and the instantaneous positions are close to corresponding value that is defined by the theoretical equation. It confirms the accuracy of water capillarity prediction by molecular dynamics method.

After the validating the water penetration process with classic capillary theory, it is worth investigating more complicated ions invasion phenomenon in the nanometer channel. Considering that the number of ions confined in the gel pore is quite small, the average penetration depth was utilized to keep the statistical stability. The mean penetration depth, defined as the mass center of ions and water molecules in the nanopore, were calculated as the function of time and shown in Fig. 5.19 Similar to

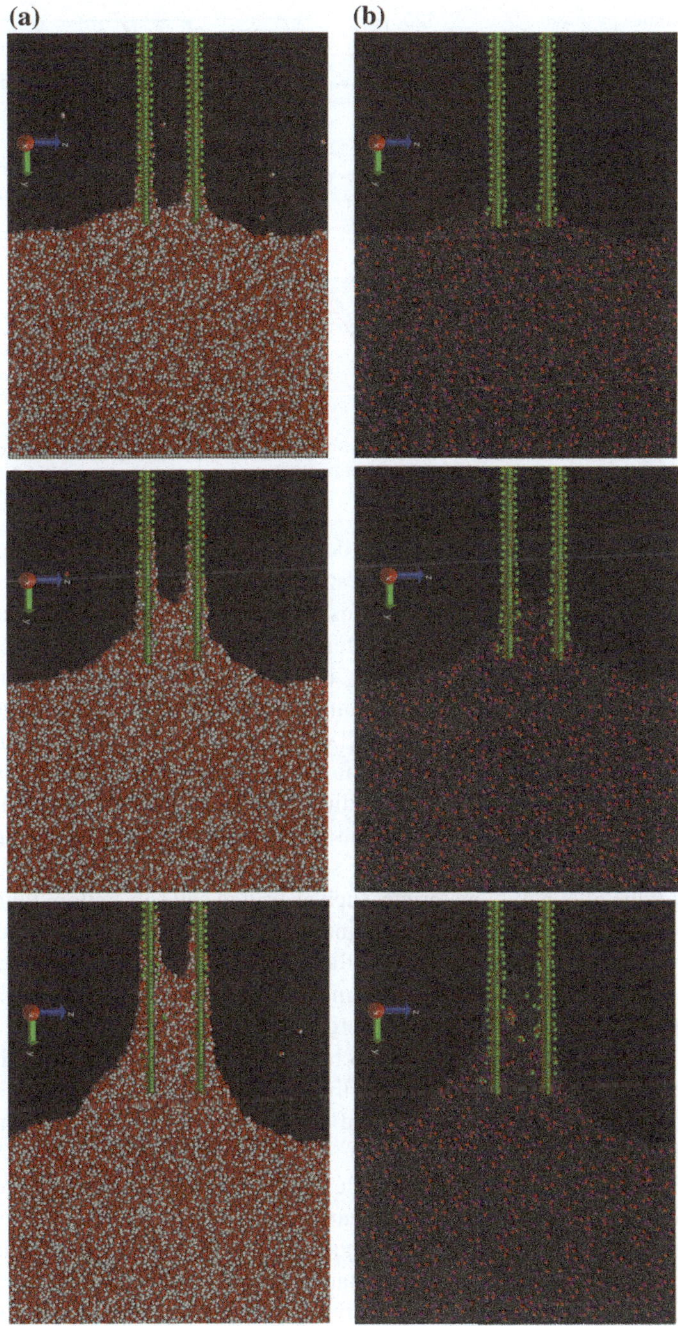

Fig. 5.17 Snapshots of capillary flow of **a** water and **b** sodium and chloride ions in the 4.5 nm pore at 100, 500, and 2000 ps (from top to bottom)

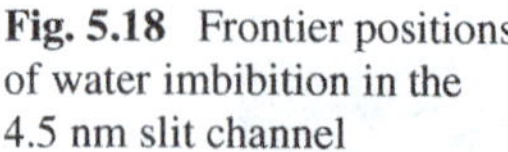

Fig. 5.18 Frontier positions of water imbibition in the 4.5 nm slit channel

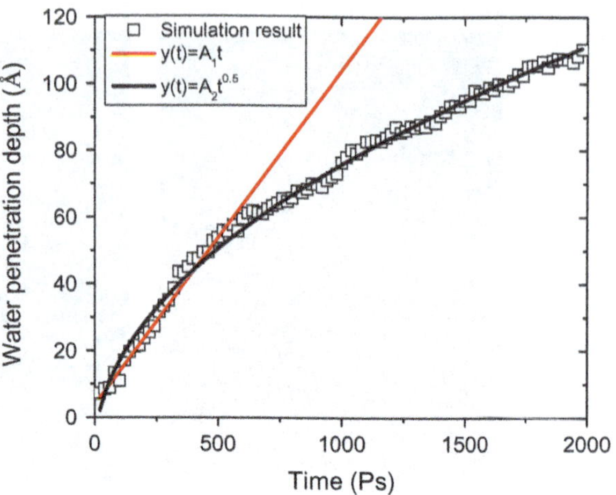

water transport, the ionic imbibition also demonstrates the transition from constant velocity regime to visco-inertia regime. Water molecules play a role in carrying ions during the transport, so the mobility of ions is significantly influenced by neighboring water molecules. Nevertheless, consistent with the observation in Fig. 5.17b, the penetration depth of both Na and Cl ions is smaller than that of water molecules. While chloride ions closely follow the waterfront, sodium ions have already traveled 20 Å behind water molecules after 2000 Ps. It means that nanometer channel has strong selectivity in different species in the solution. Due to the filtering role of C–S–H nanostructure, the chloride and sodium ions are gradually separated from the water molecules. The capillary adsorption mechanism that ions immigrate slower than water molecules is of importance for studying the durability of concrete materials. The service of concrete materials, such as bridge and sub-bottom tunnel in the marine environment, is shortened by all kinds of damage, and most of these deterioration mechanisms are strongly related to the existence and movement of water [38]. For example, water plays a role in carrying chloride ions into the pore structures of concrete, which causes the steel reinforcement corrosion and concrete structure damage [39]. The capillary adsorption mechanism of water and ions can help predict accurately the invasion process of the detrimental ions in concrete. Accordingly, the protection measures, such as a waterproof agent, can be taken to inhibit the water penetration [40].

Figure 5.20 exhibits the temporal evolution of y-direction intensity profiles for water and ions. The intensity evolution clearly describes that water and chloride ions gradually penetrate into the nanometer channel. As shown in Fig. 5.20a, the density profile can distinguish water into three states: bulk solution water, gel pore water, and water in advancing front. The density of water transport into the gel pore is around 0.8 g/cm^3, which is quite lower than that of the bulk solution. In particular, the water molecules in the advancing frontier have density even lower than 0.6 g/cm^3. With increasing of time, the gel pore water shows some local high-density regions

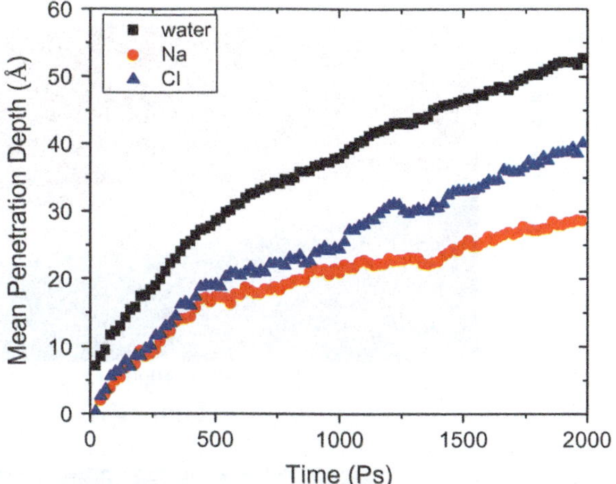

Fig. 5.19 Mean penetration depth of water, Na and Cl ions evolution with time

that are highlighted in Fig. 5.20a, indicating the condensation process of penetrated water molecules. Furthermore, it is also demonstrated in Fig. 5.20b that chloride ions show some high-intensity regions in the gel pore, which are mainly attributed to the ionic cluster formation. The y-direction position of the cluster remains almost constant from 800 ps to 1800 ps, implying that ions, aggregating together, have weak mobility. It should be noted that the high-density region of water also locates near the ionic cluster. This might give an explanation for the gel pore water transferring to high density: the ionic cluster blocks the narrow nanometer channel, further slowing down the water transport process and contributing to water condensation. The pronounced discrepancy between ions and water molecules transport can be explained by different hydrolytic diameters of atoms, nanometer confinements, and the electronic interactions with calcium silicate channel. This will be further discussed in the following section.

5.3.3 Local Structure of Water and Ions in the Gel Pore

Figure 5.21 demonstrates the molecular configuration of confined fluid and the corresponding atomic intensity profiles of water and ions normal to the calcium silicate layer. As shown in Fig. 5.21b, the intensity peaks of Ca and H_o at the interface characterize the adsorbed calcium atoms and silicate hydroxyl groups of the tobermorite surface. The position of the hydroxyl groups was arbitrarily considered as the boundary between liquid and solid. The maxima of H_w and O_w in the intensity profile are located at 1.0 and 1.8 Å away from the calcium silicate substrate. The sharp intensity peaks of water molecules near the surface are due to multilayer capillary adsorption process in the nanometer channels: water molecules first fill and form one layer close to surface, and subsequently, other molecules pack on the first

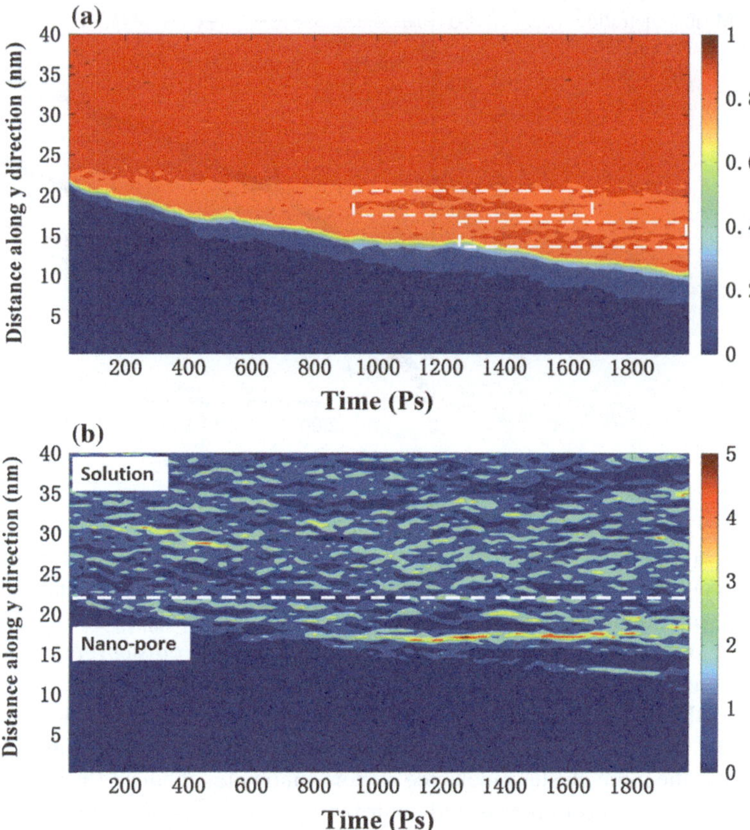

Fig. 5.20 *y*-direction (the capillary advancing direction) atomic intensity profile evolution with time. **a** Water density; **b** chloride ions number

water layer. Within the distance 10 Å from the first water layer, the water intensity gradually decreases, which contributes to the advancing menisci during the capillary adsorption. The intensity difference between H_w and O_w also indicates that the surface water molecules have the hydrogen atoms tilt toward the calcium silicate layers by forming the hydrogen bonds with oxygen atoms in silicate chains. It also confirms the hydrophilicity of the tobermorite surface.

In the ionic intensity profile, the sharp peak of sodium ions with high intensity is located 2 Å away from the surficial hydroxyls, while the relative smaller peak of chloride ions is distributed 4 Å away. It implies that as compared with chloride ions, the sodium ions are more probably to adsorb and penetrate deeper in the calcium silicate sheets. Part of the invaded ions is fixed in the cavity in the silicate chains. That is why the sodium ions transport slower in the gel pore than that of chloride ions as illustrated in Fig. 5.19 The different adsorption mechanisms for sodium and chloride ions have been shown in Fig. 5.22b and c, respectively. The sodium ions diffuse

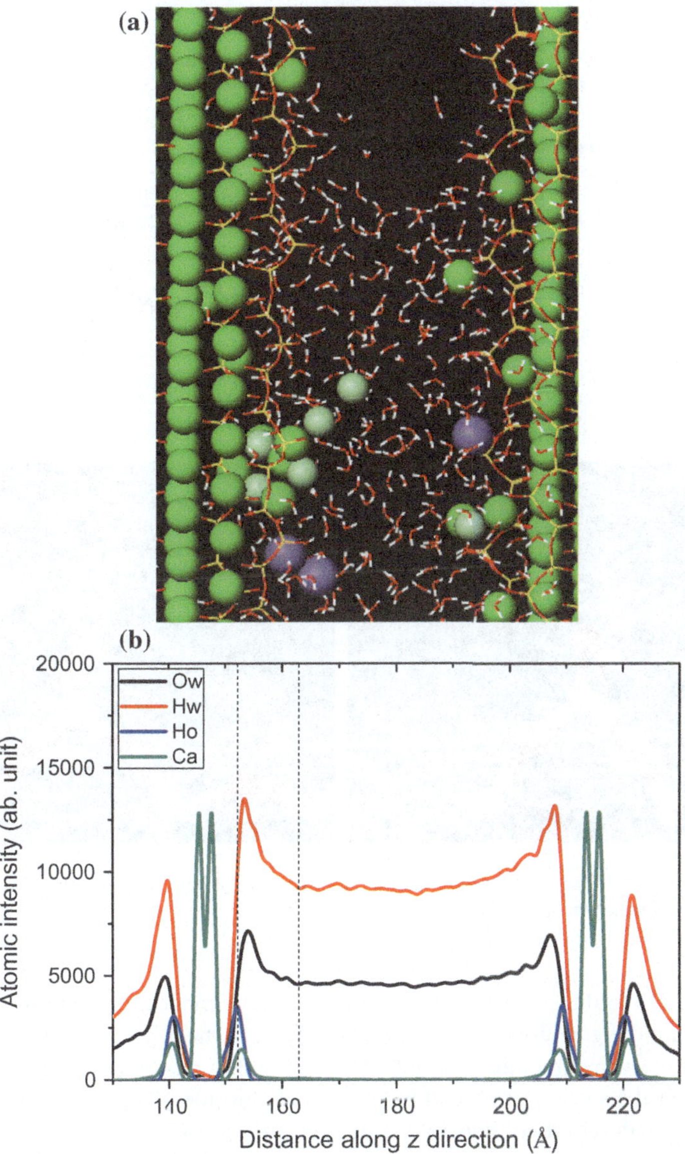

Fig. 5.21 **a** Snapshot of water and Na, Cl ions confined in the gel pore; **b** intensity distribution perpendicular to the nanometer channel (Ca, H_o, O_w, and H_w)

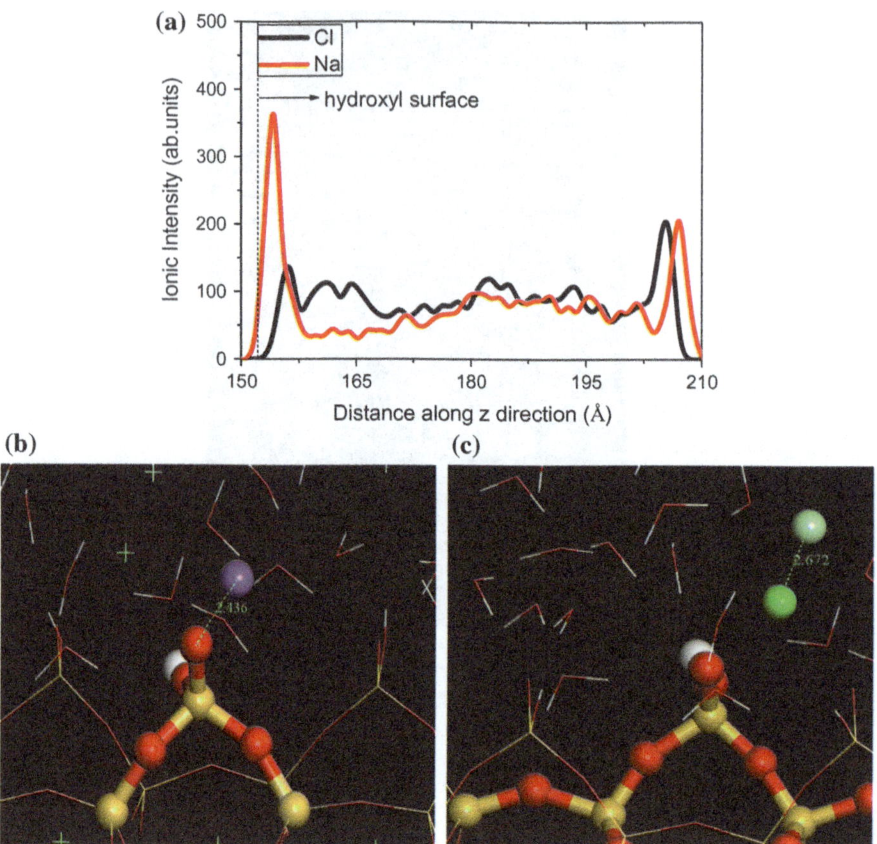

Fig. 5.22 **a** Na and Cl ions intensity profile; **b** local structure of Na ions adsorption on tobermorite surface; **c** local structure of Ca–Cl cluster near the surface

approximately to the silicate chains and form Na–O connections with neighboring non-bridging oxygen atoms in silicate hydroxyl. On the other hand, the interfacial calcium atoms can attract the chloride ions in the solution, contributing to the Ca–Cl clusters near the surface. It should be noted that chloride binding is also significant to the study of durability of concrete structures for two reasons: reduction of the free chloride concentration near the reinforcing steel so that it is not probably to induce corrosion; chloride ions are removed from the diffusion flux which can retard the imbibition of chloride to the level of the steel.

5.3.4 Dynamic Properties of Atoms in the Gel Pore

Figure 5.23a exhibits the temporal MSD variation and the y- and z-direction components of all the water molecules. The relative higher MSD values along y-direction are mainly attributed to the rapid movement along y-direction induced by the nanometer channel. The MSD components of the water molecules transport in the gel pore have been further investigated. As shown in Fig. 5.23b, after 200 ps, while the MSD_y gradually deviates to higher value, the component along z-direction remains almost unchanged. The pronounced anisotropic mobility of water confined in gel pore is partly due to strong electronic parallel sheets forcing the solvent capillary advancing along y-direction, partly due to the geometric restriction along z-direction. It should be noted that the MSD_y and MSD_{xyz} values are dramatically increased, which both are one order of magnitude higher than the bulk solution value obtained in previous simulation [35]. It means that at the very beginning of capillary adsorption, the water molecules, transport in the calcium silicate nanopore, are accelerated to a great extent so that the dynamic nature of bulk water has changed significantly. This can be interpreted as the result of strong coulombic and van der Waals dragging force as the solution and electronic C–S–H layer is contact at the beginning time.

Higher MSD_y values in Fig. 5.24a, b indicate that the negative and positive ions filling in the silicate channel also have predominately high transport rate along y-direction. It implies that the silicate channel plays a critical role in ions' acceleration in the transport process.

The atomistic displacement profile can provide more insights into the water and ions transport mechanism in the nanometer pore. As shown in Fig. 5.25, the average displacement of both ions and water molecule in the pore is more than 60 Å after only 200 ps, which is extremely higher than that of bulk courter part (around 20 Å). It implies that the calcium silicate nanopore plays a significant role in accelerating the capillary filling process. The C–S–H sheets change the random diffusion behavior of fluid and allow the solution transport with directionality. Unexpectedly, the largest displacement domain is distributed closely to the inner surface of the silicate channel, as exhibited in Fig. 5.25a. It is contrary to many observations of saturated transport in nanometer channel that water molecules ultra-confined in the nanopore or approximating to the hydrophilic surface are strongly restricted by the electronic and geometric constraints, having rather slow mobility as compared to bulk courter part [41–44]. The displacement discrepancy between saturated and unsaturated transport is mainly attributed to different local structures and pair dynamics of water molecules. As shown in Fig. 5.26a, the water molecules in the capillary advancing front region have weak chemical connections with neighboring atoms. In particular, lots of molecules, approximating to the silicate chains, are only bonded by donating H-bonds to non-bridging oxygen atoms. Once the H-bonds are broken, the water molecule can escape from the silicate surface quickly. On the other hand, in the saturated state, besides the H-bonds from the silicate interface, the surface water molecules are restricted by H-bonds contributed from neighboring 3 or 4

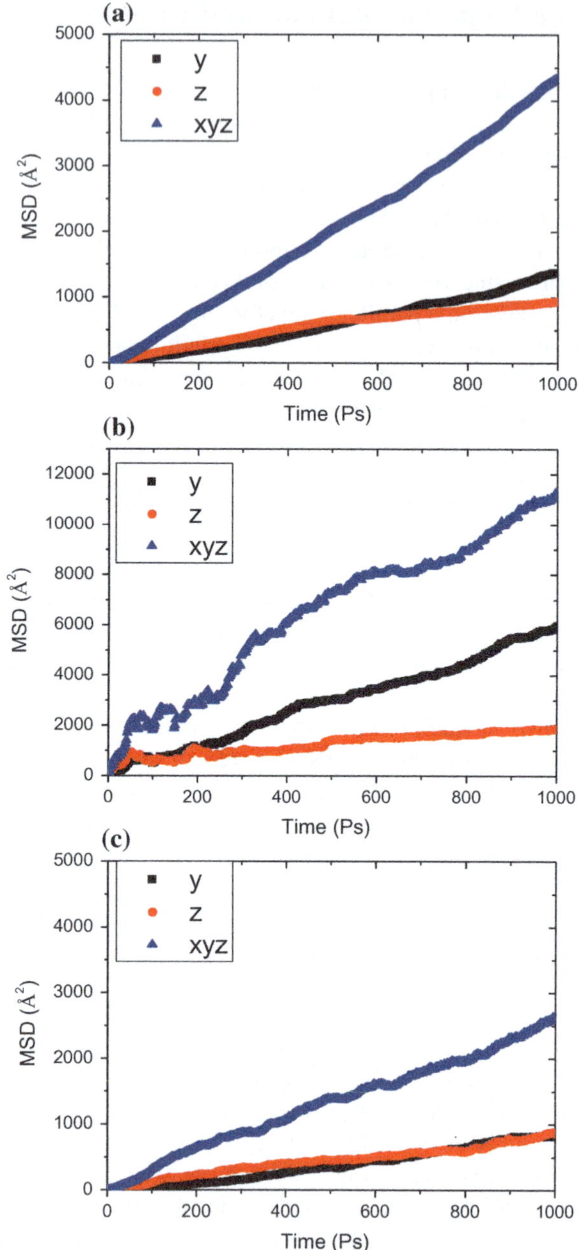

Fig. 5.23 MSD evolution during 1000 ps for **a** all the water molecules; **b** water molecules confined in the gel pore; **c** water molecules 100 Å away from the gel pore along y-direction. MSD_y and MSD_z calculate the displacement along y- and z-direction, respectively, and MSD_{xyz} is the summation of MSD_x, MSD_y and MSD_z

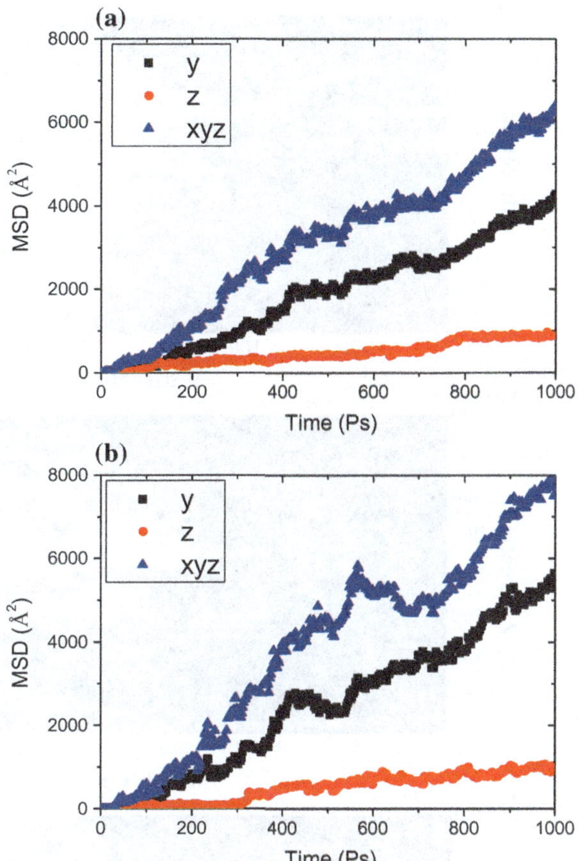

Fig. 5.24 MSD evolution for **a** chloride ions confined in the gel pore; **b** sodium ions confined in the gel pore

other molecules. These surrounding atoms can construct a "cage" for central water molecule, slowing down the diffusion process.

Nevertheless, ionic displacement profiles in Fig. 5.25b, c only show the highest mobility region in the central of the gel pore rather than near the inner surface as exhibited in the water profile in Fig. 5.25a. In the respect of local structures, it is clearly observed in Fig. 5.26b that both the chloride and sodium ions are coordinated with 6–8 neighboring atoms, no matter they approximate to or distribute away from the silicate surface. Unlike water molecules, the ions move forward by carrying the surrounding water molecules, but cannot escape from the bulk water solution and diffuse separately.

The time correlated function (TCF) is utilized to describe dynamical properties of ions–water and ions–ions correlations, as well as the bonding stability between solution species and atoms in the C–S–H gel. The border of the hydration shell is defined as the first minimum of ion–oxygen RDF. For the species in the solution, the meaning of **TCF** is the probability that a water molecule or ion, which was in the

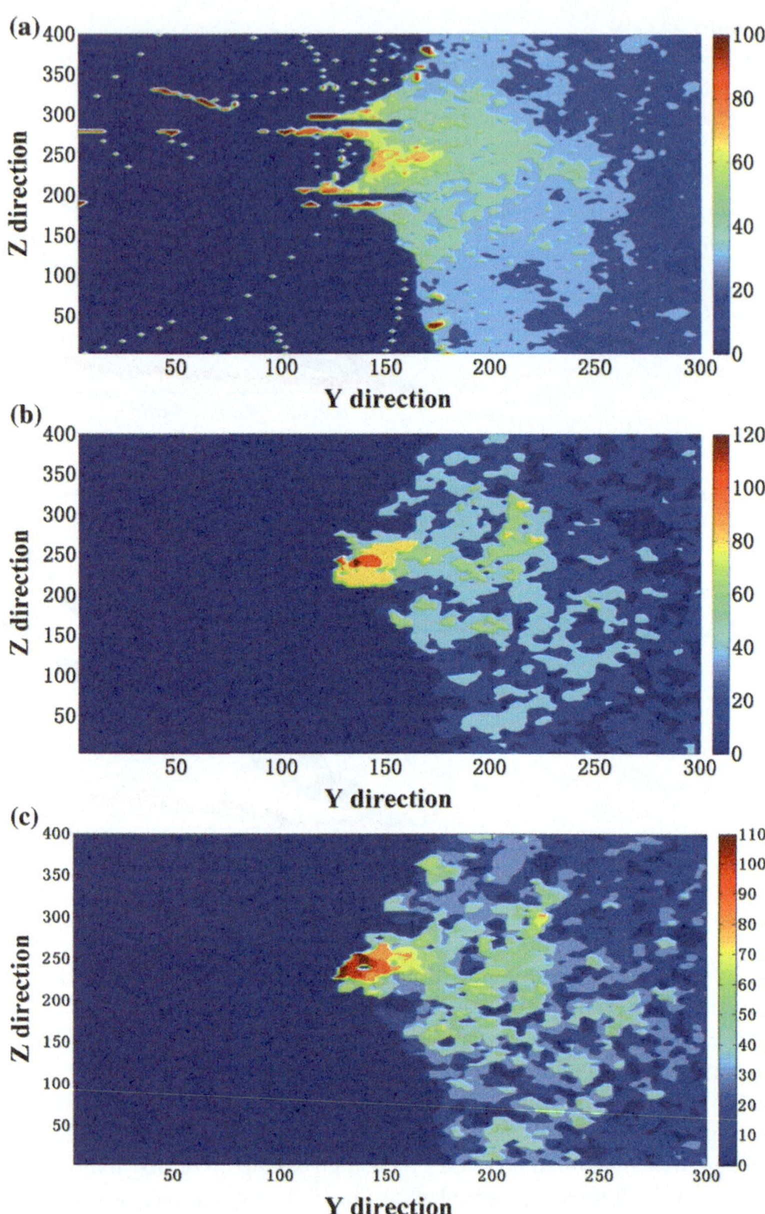

Fig. 5.25 Atomistic displacement distribution in *YZ* plane for **a** water molecule; **b** chloride ion; **c** sodium ion in 200 ps; the color map represents absolute value of displacement (unit: Å)

(a)

(b)

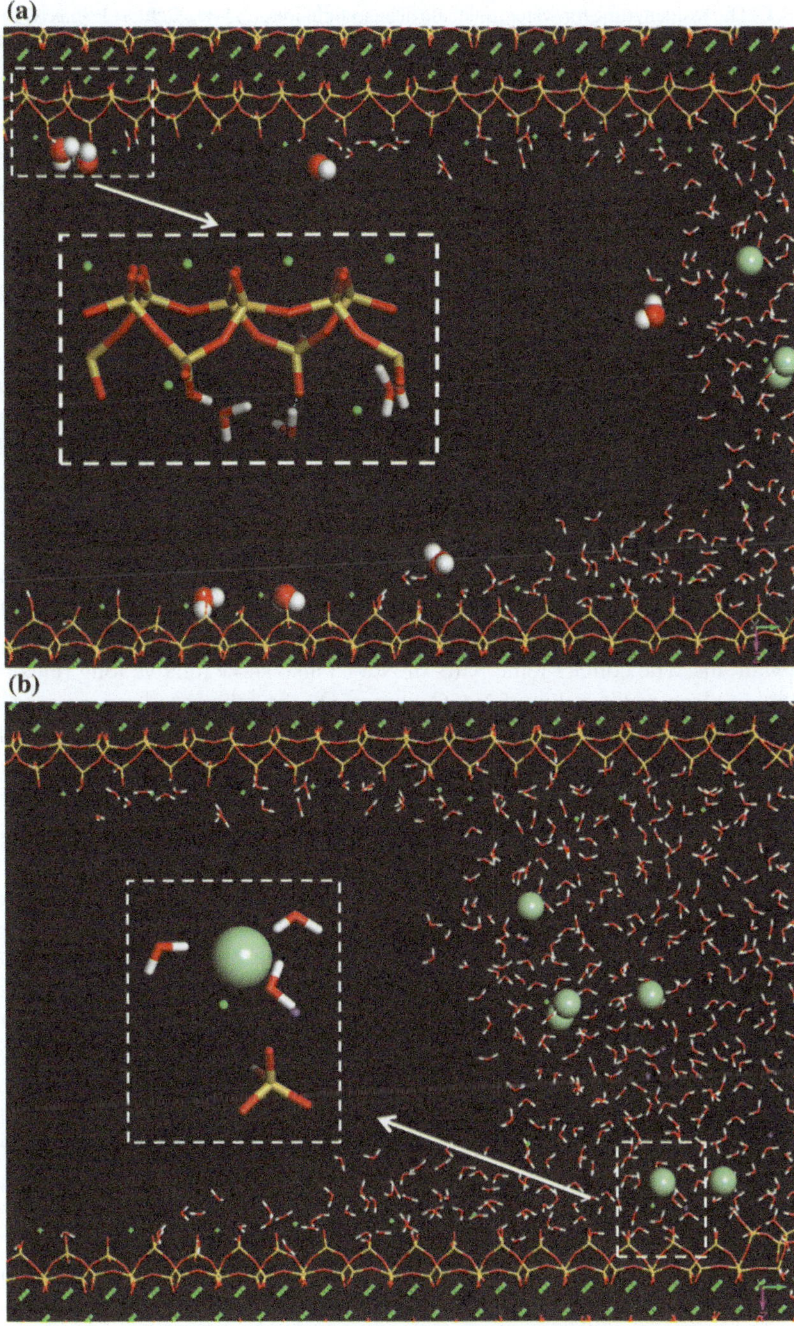

Fig. 5.26 **a** Local structure of water molecule near the silicate surface; **b** local structure of chloride ions near the silicate surface

Table 5.1 The resident time for water molecule near sodium ions ($Na-O_w$), chloride ions ($Cl-O_w$), and water molecule (O_w-O_w), the resident time for chloride ions near sodium ions ($Na-Cl$) in solution (τ_{sol}) and in gel pore (τ_{pore}); relaxation time of different bonds formed between water and ions near the surface (τ_{surf})

	$Na-O_w$	$Cl-O_w$	O_w-O_w	$Na-Cl$
τ_{sol} (Ps)	38.71	14.30	4.50	32.48
τ_{pore} (Ps)	35.93	13.65	4.28	33.49
	O_w-O_s	H_s-O_w	$Ca-Cl$	$Na-O_s$
τ_{surf} (Ps)	26.17	25.62	>168.83	>131.63

hydration shell of ion i initially, is also in the hydration shell of the same at time t. TCF for the solution species and surface atoms indicates that a molecule or ion, which formed chemical bonds with surface atom i at the very beginning, remains binding with the same atom at time i. The TCF for ions–water and ions–ions, ions–surface are exhibited in Fig. 5.27.

Also, the values of residence time are listed in Table 5.1. It is clearly observed in Table 5.1 that both in the gel pore and solution, the relaxation time for water near the sodium ions is more than twice longer than that for water near the chloride ions. It implies that the short ions–water distance contributes to the strong interaction between atoms. The contact ionic pairs of Na–Cl also show relative long relaxation time that is close to the value of $Na-O_w$ pairs. On the other hand, on average, the nearest neighboring molecules can only remain near the central water molecule in less than 5 ps, no matter in bulk solution and in nanometer channel. The short relaxation time of water–water pairs is mainly attributed to frequently breakage of the H-bonds between water molecules. The large resident time discrepancy between ions–water and water–water provides interpretation on different transport mechanism for ions and water molecules. It is difficult for the sodium ions to diffuse away from the hydration shell constructed by the $Na-O_w$ with strong bond strength. That is why chloride and sodium ions carry plenty of water molecules, forming large ions–water cluster, as they transport in nanometer pore. Due to the presence of negatively charged silicate chains, the water molecules energetically prefer to escape from the hydration cage by the weak H-bonds.

Furthermore, the TCF between water (ions) and atoms in the calcium silicate sheet can reflect the resident time of water and ions near the channel surface. According to the pair delaying rate in **b**, chemical bonds are ranked in the following order by comparing the bond strength: $Ca-Cl > Na-O_s \gg$ H-bonds. As compared with water molecules, the chloride and sodium ions can form a stable connection with the surface calcium ions and non-bridging oxygen atoms in the silicate chain, remaining in the vicinity of tobermorite surface for a longer time. The relative slower transport rate of Na and Cl ions is mainly attributed to the pronounced stabilizing effect from the calcium silicate layers. Additionally, the long relaxation time of Na–Cl and Ca–Cl contact pairs indicates the chloride ions carry lots of positive ions, as transport in the gel pore, slowing down the ions imbibition in the nanometer channel. With the

Fig. 5.27 Time correlated function for ions–ions and ions–water pair **a** in gel pore; **b** in solution; **c** TCF for water and surface atoms, ions and surface atoms

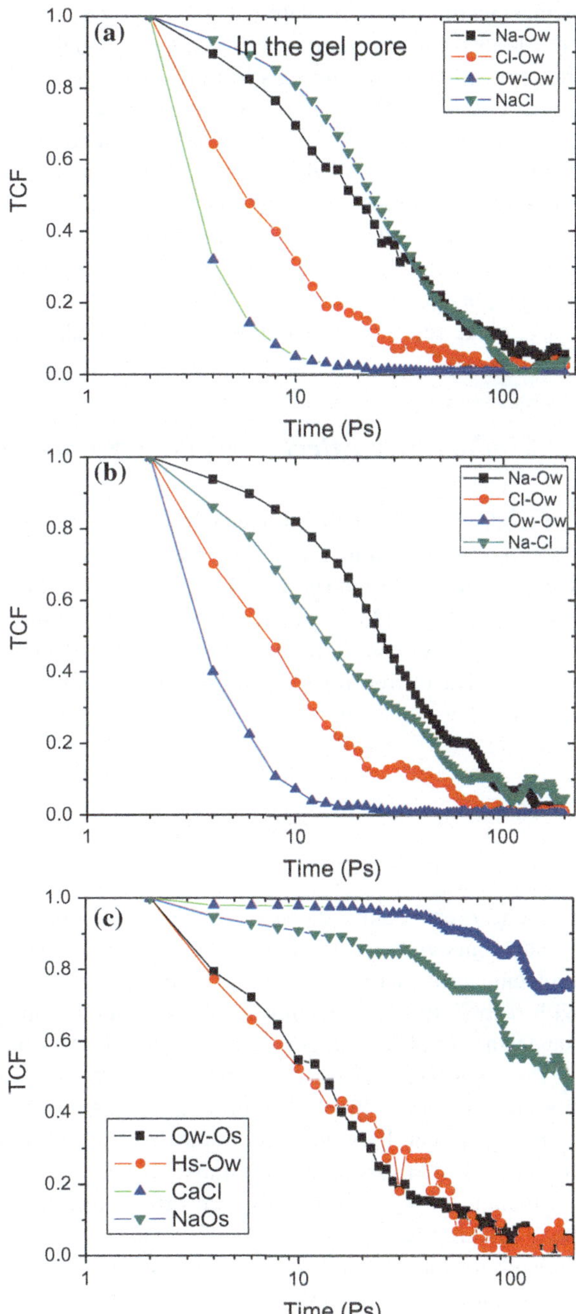

increasing number of penetrated ions, the contact Ca–Cl pair can grow to large Ca–Cl–Na cluster with more ions aggregate together, which blocks the nanochannel and prevents solvents and ions from further penetrating.

More importantly, the mechanism of immobilization of calcium silicate hydrate on the cation ions can help understand the interaction of metal ions with the cement matrix. It is valuable that the cation exchange and cluster aggregation mechanism are applied to investigate the immobilization of radioactive waste by the cement matrix. A better understanding of the immobilization mechanism can guide the cementitious materials design in the near field of the radioactive waste repository [45–47]. For example, the cement matrix can be utilized as shield cask and grouting for sealing cracks in the repository to inhibit the nuclear radiation from waste.

5.3.5 Pore Size Effect on Capillary Transport

In order to evaluate the influence of pore size on the capillary adsorption process, 6 and 3.5 nm calcium silicate pores were constructed and simulated and analyzed with same method as the case of 4 nm. The mean penetration depth of water and ions is shown in Fig. 5.28. In the constant velocity regime at the very beginning from 0 to 250 ps, the penetration rate for the water molecules reduces with increasing gel pore size. The decreasing imbibition velocity for increasing channel length is consistent with theoretical study proposed by Bosanquet [48], as defined in the following equation:

$$y(t) = At = \sqrt{\frac{\sigma \cos \theta}{\rho r}} \cdot t \tag{5.2}$$

where σ, θ and r have the same meaning as mentioned in Eq. (5.1), ρ is the liquid density. This equation explains that due to the weaker viscosity force, the inertial drag and capillary force contribute to the accelerating of fluid at the very beginning. With respect to the molecular level, the calcium silicate sheets with narrower distance can provide a stronger electronic field to drag water molecules in the channel. On the other hand, the visco-inertia regime after 800 ps shows that water molecules migrate faster in channel of 6 nm than that of 4.5 and 3.5 nm. It means that at the later stage of capillary adsorption, the viscosity plays critical role in slowing down the transport. At this stage, the transport velocity reduction in smaller gel pore is significant due to more pronounced effect from the geometrical confinement and the elongating resident time at C–S–H surface. It is worth noting that the chloride and sodium ions migration also demonstrates opposite penetration depth and pore size relation for water molecules at different capillary stage. Furthermore, it can be observed in Fig. 5.28b, c that the imbibition velocity of ions reduces much faster than that of water molecules in the channel of 3.5 and 4.5 nm. It means that the viscous role from nanoconfinement and calcium silicate surface influences the ionic trans-

Fig. 5.28 Penetration depth of **a** water molecules; **b** chloride ions; **c** sodium ions as a function of time in nanopore with diameter 6, 4.5 and 3.5 nm

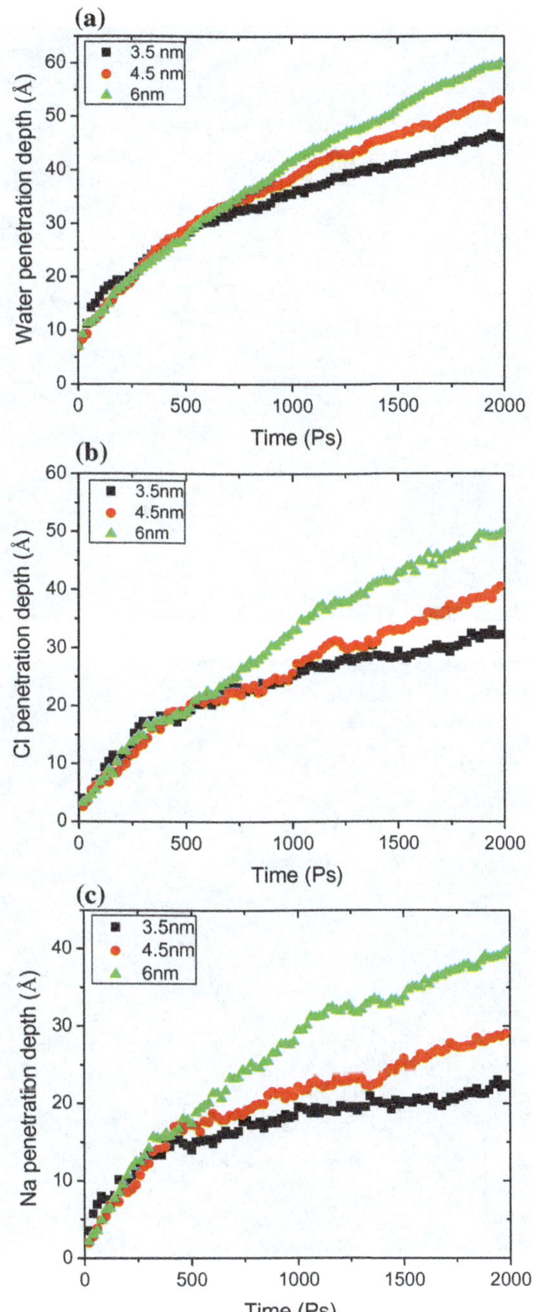

(a)

(b)

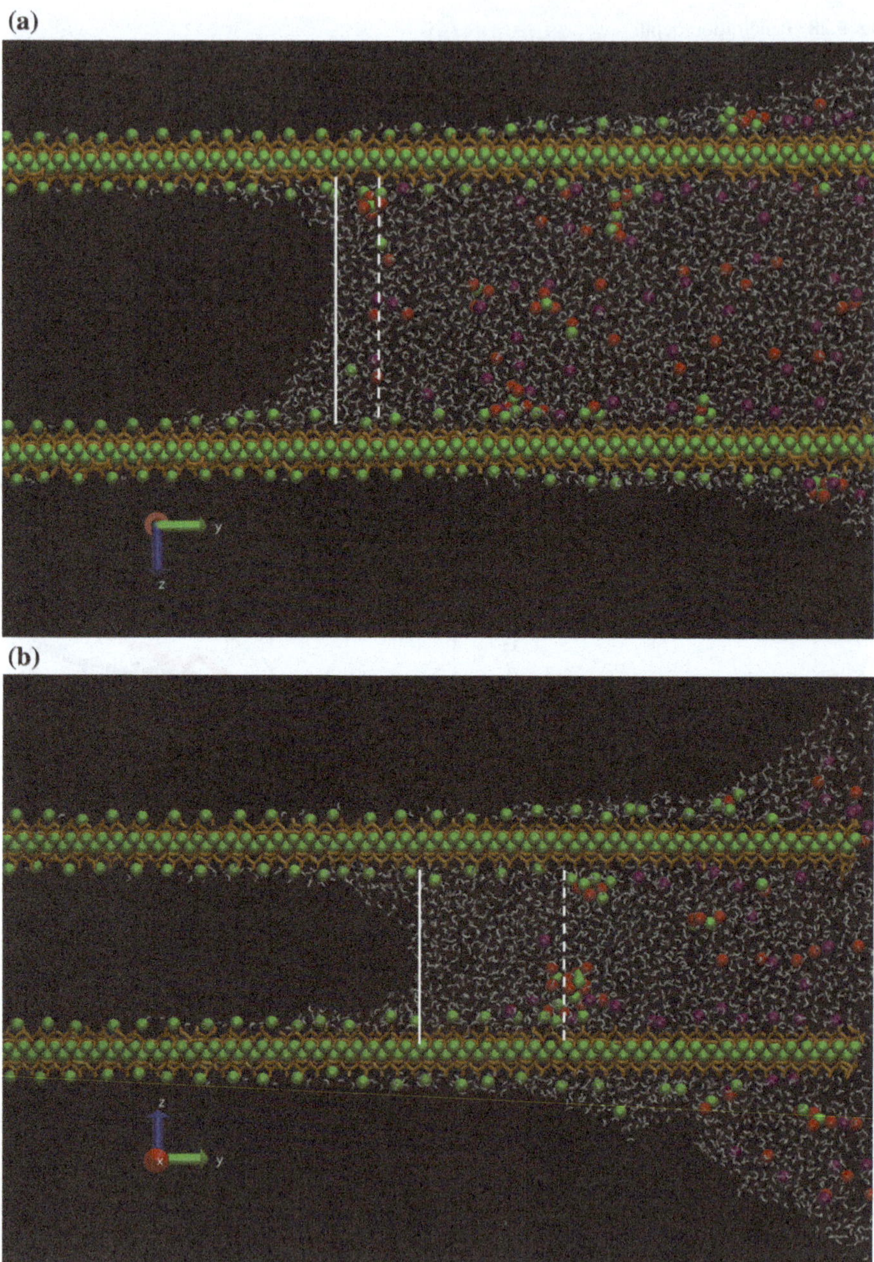

Fig. 5.29 Snapshots of water and ions penetrated in the nanometer channel with diameter of **a** 6 nm; **b** 4.5 nm; **c** 3.5 nm at 2000 ps. The white real line represents the water frontier and the dash line is the chloride ions frontier

(c)

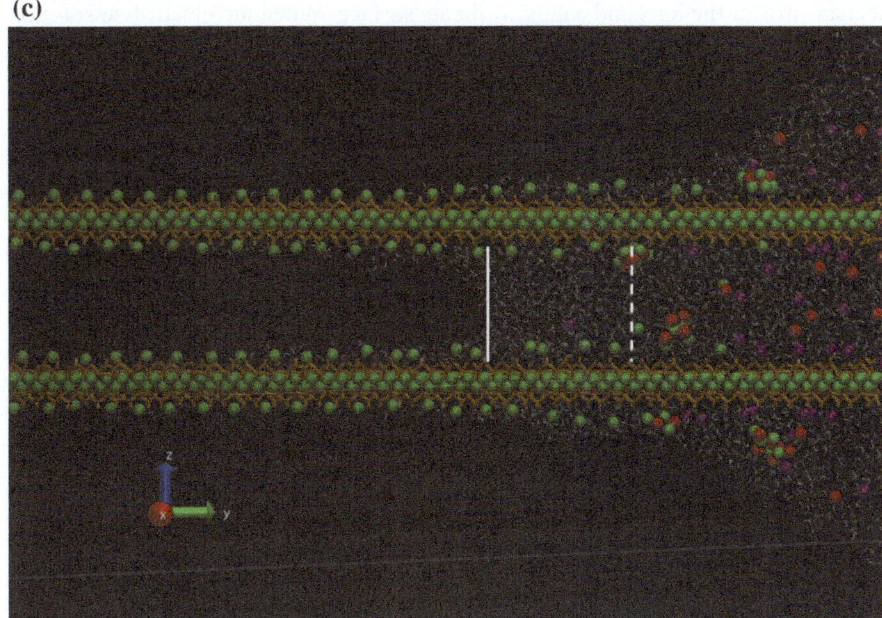

Fig. 5.29 (continued)

port more significantly. Previous experimental studies have found that the chloride diffusivity and water permeability greatly depend on the pore structure of the cement and concrete matrix [49, 50]. Hence, in marine concrete design, critical pore size is considered as the most important parameter to estimate the durability of concrete material [51]. Controlling the water/cement ratio, increasing curing age and incorporating functional additives are effective measurements to improve the pore structure of cement matrix, which can further retard the penetration rate of water and chloride [52].

Figure 5.29 exhibits the configurations of water and ions in the nanometer channel at 2000 ps. It can be clearly observed that the distance between the capillary frontier of water molecules and ions turns larger with reducing pore size. It implies that the ionic migration is more likely to separate from the water penetration in the narrow channel. The transport discrepancy between water and ions is more pronounced with the decreasing of nanopore size due to the following reasons: (a) The small nanometer channel plays filtering role to prevent from entrancing the chloride and sodium ions with larger hydration shell; (b) the ionic pairs such as Ca^{2+} and Cl^- are more likely to accumulate to form cluster, blocking further penetration of other ions; (c) the immobilization effect is quite stronger from the surface calcium ions and non-bridging oxygen atoms to elongate the resident time of ions.

It can also be observed that the transport velocity of sodium ions is slower than that of chloride ions. Behind the ionic transport, the discrepancy is the different interaction mechanisms for the ions and calcium silicate surface. When the electrolyte solution flows through calcium silicate channel, sodium ions associate with the oxygen atoms in the silicate chains, substituting the surface calcium ions. It is consistent with the experimental evidence that the alkaline ions can reside both in the interlayer region and surface of calcium silicate layers by the cation exchange, as the C–S–H gel is subjected to the alkaline solution [53]. Due to the strong electronic attraction from silicate chains, the sodium ions are immobilized near the surface for a long time. The ionic exchange results in the calcium atoms dissociating from the silicate surface and diffusing in the gel pore solution. Calcium ions in the aqueous state can capture the neighboring chloride ions, aggregating to the Ca–Cl cluster. With the accumulation of the cluster, the transport velocity of chloride ions is further reduced.

5.4 Chapter Summary

In this chapter, Molecular dynamics was utilized to model water and ions confined in the nanometer channel of C–S–H mineral and the capillary transport of Na^+, Cl^- ions and water molecules in the nanometer channel constructed by calcium silicate hydrate. The molecular structure, dynamics of ions and water molecules, and solution–substrate interaction were analyzed to make the following conclusions:

(1) The silicate channel geometry and the interfacial negative charge make the channel water have special structural and dynamic features: high density, orientation preference, ordered interfacial organization, and low diffusion rates. In addition, the channel water molecules, defined as type 1 and type 2 water, have more H-bonds connections contributed by the substrate structures. It is worth noting that type 1 and type 2 water molecules demonstrate similarity with the interlayer water in the layered C–S–H gels, both structurally and dynamically.

(2) In addition to the surface water feature, the ion interaction in the tobermorite pores has also been investigated. Due to the negative interfacial charge, tobermorite prefers repulsing Cl^- and adsorbing Ca^{2+}. On the contrary, the surface Cl^-, cannot develop a stable connection with the tobermorite substrate. A binding of less than 5% of Cl^- binding near the surface in our simulation confirms the weak Cl^- sorption ability of tobermorite that was stated in the experimental results achieved in ^{35}Cl NMR tests. The Cl^- ions can form clusters with the surface adsorbed Ca^{2+}, but the Ca–Cl pair is not stable in the interfacial region. For long timescale simulation (larger than 1 ns), chloride ions disturb the stability for the surface adsorbed calcium atoms and most of the surface calcium atoms diffuse from the silicate channel and move away from the original position.

(3) During the capillary transport process, the early stage of water imbibition depth in the C–S–H gel as the function time includes the initial constant velocity regime of linear temporal relation, followed by visco-inertia regime with parabolic time relation, which agrees well with classic capillary adsorption theory and the experimental observation. With increasing pore size from 3.5 to 6 nm, the initial

constant velocity reduces and the penetration depth increases for both ions and water molecules.

(4) Dynamically, the diffusion coefficient of water and ions in the transport frontier is dramatically increased as compared with the value of the bulk solution. This accelerating effect is attributed to the predominant role of the inertia dragging force from electronic charged C–S–H surface. The displacement profiles show the highest mobility region in the center of the gel pore for ions and near the inner surface for confined water molecules.

(5) The water and ions have different capillary transport behavior: the ions migrate slower than water molecules in the C–S–H gel pore. Molecular dynamics attributes the different transport mechanism of water and ions in porous cementitious materials to three reasons: (a) filtering effect is pronounced for the nanostructure of hydration product and nanometer channel screens chloride and sodium ions with larger hydration shell; (b) as compared with water, ions reside in the calcium silicate surface for longer time due to the strong chemical bonding from surface silicate chains and calcium atoms; (c) chloride ions are more probable to form CaCl ionic cluster that accumulates in the nanometer channel and slows down the migrating rate.

References

1. Yu, J., Yool, A. J., Schulten, K., & Tajkhorshid, E. (2006). Mechanism of gating and ion conductivity of a possible tetrameric pore in aquaporin-1. *Structure, 14*(9), 1411–1423.
2. Gong, X., Li, J., Xu, K., Wang, J., & Yang, H. (2010). A controllable molecular sieve for Na^+ and K^+ ions. *Journal of the American Chemical Society, 132*(6), 1873–1877.
3. Liu, Q. F., Yang, J., Xia, J., Easterbrook, D., Li, L. Y., & Lu, X. Y. (2015). A numerical study on chloride migration in cracked concrete using multi-component ionic transport models. *Computational Materials Science, 99*, 396–416.
4. Günes, S., & Sariciftci, N. S. (2008). Hybrid solar cells. *Inorganica Chimica Acta, 361*(3), 581–588.
5. Jennings, H. M. (2008). Refinements to colloid model of CSH in cement: CM-II. *Cement and Concrete Research, 38*(3), 275–289.
6. McDonald, P. J., Korb, J. P., Mitchell, J., & Monteilhet, L. (2005). Surface relaxation and chemical exchange in hydrating cement pastes: A two-dimensional NMR relaxation study. *Physical Review E, 72*(1), 011409.
7. Thomas, J. J., FitzGerald, S. A., Neumann, D. A., & Livingston, R. A. (2001). State of water in hydrating tricalcium silicate and Portland cement pastes as measured by quasi-elastic neutron scattering. *Journal of the American Ceramic Society, 84*(8), 1811–1816.
8. Peterson, V. K., Neumann, D. A., & Livingston, R. A. (2005). Hydration of tricalcium and dicalcium silicate mixtures studied using quasielastic neutron scattering. *The Journal of Physical Chemistry B, 109*(30), 14449–14453.
9. Bordallo, H. N., Aldridge, L. P., & Desmedt, A. (2006). Water dynamics in hardened ordinary Portland cement paste or concrete: From quasielastic neutron scattering. *Journal of Physical Chemistry B, 110*, 17966–17976.
10. Cerveny, S., Arrese-Igor, S., Dolado, J., Gaitero, J., Alegria, A., & Colmenero, J. (2011). Effect of hydration on the dielectric properties of C–S–H gel. *Journal of Physical Chemistry, 134*, 034509.

11. Korb, J. P., Mcdonald, P. J., Monteilhet, L., Kalinichev, A. G., & Kirkpatrick, R. J. (2007). Comparison of proton field-cycling relaxometry and molecular dynamics simulations for proton–water surface dynamics in cement-based materials. *Cement and Concrete Research, 37,* 348–350.

12. Romero-Vargas Castrillón, S., Giovambattista, N., Aksay, I. A., & Debenedetti, P. G. (2009). Effect of surface polarity on the structure and dynamics of water in nanoscale confinement. *The Journal of Physical Chemistry B, 113*(5), 1438–1446.

13. Lombardo, T. G., Giovambattista, N., & Debenedetti, P. G. (2009). Structural and mechanical properties of glassy water in nanoscale confinement. *Faraday Discussions, 141,* 359–376.

14. Siboulet, B., Molina, J., Coasne, B., Turq, P., & Dufreche, J. F. (2013). Water self-diffusion at the surface of silica glasses: Effect of hydrophilic to hydrophobic transition. *Molecular Physics, 111*(22–23), 3410–3417.

15. Dufrêche, J. F., Marry, V., Malikova, N., & Turq, P. (2005). Molecular hydrodynamics for electro-osmosis in clays: From Kubo to Smoluchowski. *Journal of Molecular Liquids, 118*(1), 145–153.

16. Puibasset, J., & Pellenq, R. M. (2003). Confinement effect on thermodynamic and structural properties of water in hydrophilic mesoporous silica. *The European Physical Journal E, 12*(1), 67–70.

17. Puibasset, J., & Pellenq, R. J. M. (2004). A comparison of water adsorption on ordered and disordered silica substrates. *Physical Chemistry Chemical Physics, 6*(8), 1933–1937.

18. Rotenberg, B., Marry, V., Vuilleumier, R., Malikova, N., Simon, C., & Turq, P. (2007). Water and ions in clays: Unraveling the interlayer/micropore exchange using molecular dynamics. *Geochimica et Cosmochimica Acta, 71*(21), 5089–5101.

19. Marry, V., Rotenberg, B., & Turq, P. (2008). Structure and dynamics of water at a clay surface from molecular dynamics simulation. *Physical Chemistry Chemical Physics, 10*(32), 4802–4813.

20. Kalinichev, A. G., & Kirkpatrick, R. J. (2002). Molecular dynamics modeling of chloride binding to the surfaces of calcium hydroxide, hydrated calcium aluminate, and calcium silicate phases. *Chemistry of Materials, 14*(8), 3539–3549.

21. Kalinichev, A. G., Wang, J., & Kirkpatrick, R. J. (2007). Molecular dynamics modeling of the structure, dynamics and energetics of mineral water interfaces: Application to cement materials. *Cement and Concrete Research, 37,* 337–347.

22. Cygan, R. T., Liang, J. J., & Kalinichev, A. G. (2004). Molecular models of hydroxide, oxyhydroxide, and clay phases and the development of a general force field. *Journal of Physical Chemistry, 108,* 1255–1266.

23. Wang, J. W., Kalinichev, A. G., & Kirkpatrick, R. J. (2004). Molecular modeling of water structure in nano-pores between brucite (001) surfaces. *Geochimica et Cosmochimica Acta, 68,* 3351–3365.

24. Hamid, S. (1981). The crystal structure of the 11 A natural tobermorite $Ca_{2.25}Si_3O_{7.5}(OH)_{1.5}H_2O$. *Zeitschrift fur Kristallographie, 154,* 189–198.

25. Murray, S. J., Subramani, V. J., Selvam, R. P., & Hall, K. D. (2010). Molecular dynamics to understand the mechanical behavior of cement paste. *Transportation Research Record: Journal of the Transportation Research Board,* 75–82.

26. Mindess, S., Young, J. F., & Darwin, D. (2003). *Concrete.* Prentice Hall PTR.

27. Manzano, H., Pellenq, R. J. M., Ulm, F. J., & Buehler, M. J. (2012). Hydration of calcium oxide surface predicted by reactive force field molecular dynamics. *Langmuir, 28,* 4187–4197.

28. Hou, D. S., & Li, Z. J. (2014). Molecular dynamics study of water and ions transported during the nanopore calcium silicate phase: Case study of jennite. *Journal of Materials in Civil Engineering, 26*(5).

29. Youssef, M., Pellenq, R. J. M., & Yildiz, B. (2011). Glassy nature of water in an ultraconfining disordered material: The case of calcium silicate hydrate. *Journal of American Chemistry Society, 133*(8), 2499–2510.

30. Coudert, F. X., Vuilleumier, R., & Boutin, A. (2006). Dipole moment, hydrogen bonding and IR spectrum of confined water. *ChemPhysChem, 7*(12), 2464–2467.
31. Korb, J. P., Monteilhet, L., McDonald, P. J., & Mitchell, J. (2007). Microstructure and texture of hydrated cement-based materials: A proton field cycling relaxometry approach. *Cement and Concrete Research, 37*(3), 295–302.
32. Fratini, E., Faraone, A., Ridi, F., Chen, S. H., & Baglioni, P. (2013). Hydration water dynamics in tricalcium silicate pastes by time-resolved incoherent elastic neutron scattering. *The Journal of Physical Chemistry C, 117*(14), 7358–7364.
33. Pan, T. Y., & Liu, Y. J. (2009). Computational molecular analysis of chloride transport in hydrated cement paste. *Transportation Research Record: Journal of the Transportation Research Board, 2113*(1), 31–40.
34. Chatterji, S., & Kawamura, M. (1992). Electrical double layer, ion transport and reactions in hardened cement paste. *Cement and Concrete Research, 22*(5), 774–782.
35. Hou, D., & Li, Z. (2014). Molecular dynamics study of water and ions transport in nano-pore of layered structure: A case study of tobermorite. *Microporous and Mesoporous Materials, 195*, 9–20.
36. Plimpton, S., Thompson, A., Crozier, P., & Kohlmeyer, A. (2001). LAMMPS molecular dynamics simulator.
37. Washburn, E. W. (1921). The dynamics of capillary flow. *Physical Review, 17*, 273–283.
38. Ma, H., Hou, D., Liu, J., & Li, Z. (2014). Estimate the relative electrical conductivity of C–S–H gel from experimental results. *Construction and Building Materials, 71*, 392–396.
39. Zhang, J., Ma, H., Pei, H., & Li, Z. (2016). Steel corrosion in magnesia–phosphate cement concrete beams. *Magazine of Concrete Research, 69*(1), 35–45.
40. Medeiros, M. H., & Helene, P. (2009). Surface treatment of reinforced concrete in marine environment: Influence on chloride diffusion coefficient and capillary water absorption. *Construction and Building Materials, 23*(3), 1476–1484.
41. Hou, D., Lu, C., Zhao, T., Zhang, P., & Ding, Q. (2015). Structural, dynamic and mechanical evolution of water confined in the nanopores of disordered calcium silicate sheets. *Microfluidics and Nanofluidics, 19*(6), 1309–1323.
42. Hou, D., Zhao, T., Ma, H., & Li, Z. (2015). Reactive molecular simulation on water confined in the nanopores of the calcium silicate hydrate gel: Structure, reactivity, and mechanical properties. *The Journal of Physical Chemistry C, 119*(3), 1346–1358.
43. Hou, D. S., & Li, Z. J. (2013). Molecular dynamics study of water and ions transport in nano-pore of calcium silicate phase: A case study of jennite. *Journal of Materials in Civil Engineering*.
44. Youssef, M., Pellenq, R. J. M., & Yildiz, B. (2011). Glassy nature of water in an ultraconfining disordered material: The case of calcium–silicate–hydrate. *Journal of the American Chemical Society, 133*(8), 2499–2510.
45. Dutré, V., & Vandecasteele, C. (1998). Immobilization mechanism of arsenic in waste solidified using cement and lime. *Environmental Science and Technology, 32*(18), 2782–2787.
46. Gougar, M. L. D., Scheetz, B. E., & Roy, D. M. (1996). Ettringite and C–S–H Portland cement phases for waste ion immobilization: A review. *Waste Management, 16*(4), 295–303.
47. El-Kamash, A. M., El-Naggar, M. R., & El-Dessouky, M. I. (2006). Immobilization of cesium and strontium radionuclides in zeolite-cement blends. *Journal of Hazardous Materials, 136*(2), 310–316.
48. Bosanquet, C. H. (1923). On the flow of liquids into capillary tubes. *Philosophical Magazine Series, 6*(45), 525–531.
49. Halamickova, P., Detwiler, R. J., Bentz, D. P., & Garboczi, E. J. (1995). Water permeability and chloride ion diffusion in Portland cement mortars: Relationship to sand content and critical pore diameter. *Cement and Concrete Research, 25*(4), 790–802.
50. Poon, C. S., Kou, S. C., & Lam, L. (2006). Compressive strength, chloride diffusivity and pore structure of high performance metakaolin and silica fume concrete. *Construction and Building Materials, 20*(10), 858–865.

51. Ma, H., & Li, Z. (2013). Realistic pore structure of Portland cement paste: Experimental study and numerical simulation. *Computer and Concrete, 11*(4), 317–336.
52. Chindaprasirt, P., Rukzon, S., & Sirivivatnanon, V. (2008). Resistance to chloride penetration of blended Portland cement mortar containing palm oil fuel ash, rice husk ash and fly ash. *Construction and Building Materials, 22*(5), 932–938.
53. Lothenbach, B., & Nonat, A. (2015). Calcium silicate hydrates: Solid and liquid phase composition. *Cement and Concrete Research, 78*, 57–70.

Chapter 6
Models for the Cross-Linked Calcium Aluminate Silicate Hydrate (C–A–S–H) Gel

How to lower the environmental footprint in the construction industry challenges the researchers in this field. The key measure should lie in the reduction of cement usage, since the manufacture of cement is a high energy cost industry that results in about 6–8% of the yearly man-made global CO_2 emissions [1]. One common practice is to prepare concrete with cement clinker partially substituted by supplementary cementitious materials (SCMs) (e.g., blast furnace slag and fly ash, are also industrial wastes). Studies suggest that SCMs addition can change the hydration product distribution in the Portland cement (PC) paste [2–4] and help it resist the sulfate attack [5, 6], hence improving the durability of concrete structure under marine environment. From another point of view, longer service life of this kind of cement-based material can also lower the carbon footprint. Another promising method is the full usage of industrial wastes combined with alkali activator to produce alkali-activated cements (AACs), where the alkali-activated slag (AAS) binder has been intensively studied and applied as an alternative to PC in many places of the world [7, 8].

6.1 Background of Cross-Linked C–A–S–H Gel

The application of SCMs, including aluminate-rich minerals, also changes the chemical stoichiometry of the cement-based materials [9]. The incorporation of some aluminum into the C–S–H contributes to the formation of C–A–S–H gel. As the C–A–S–H gel is the main binding phase in the Portland cement blended with a high volume of aluminum-rich SCM and also in AAS, its structure and properties greatly influence the performance of PC blends and AAS paste [9, 10]. Furthermore, recent research on Roman concrete showed that its main hydration product is the C–A–S–H gel [11]. This old concrete maintains its structural integrity under marine environment for over 2000 years, arising the conjecture that the high durability of this concrete is correlated with the presence of C–A–S–H gel. This also demonstrates the importance of C–A–S–H gel to the sustainability of the building industry. Many studies were carried out to investigate the molecular structure of C–A–S–H [12–18]. Some

© Science Press and Springer Nature Singapore Pte Ltd. 2020
D. Hou, *Molecular Simulation on Cement-Based Materials*,
https://doi.org/10.1007/978-981-13-8711-1_6

C–A–S–H structure models were proposed based on experimental studies [19, 20]. The majority models were constructed by modifying the well-known sandwich-like tobermorite model proposed by Taylor [21, 22] that dreierketten like defective silicate chains, including bridging and pairing tetrahedra, grow on calcium sheets to form the calcium silicate sheet, and the neighboring calcium silicate sheets are connected by the interlayer water and calcium ions. It is widely accepted that the presence of Al species in C–S–H gel is to substitute the Si atoms in the chains. While Al[4] mainly exists in the bridging site of silicate chains [23] (Al substitution can also be present in the pairing site but with higher energy penalty [15, 24]), Al in the interlayer region of calcium silicate sheets are in the form of five-/sixfold coordination, i.e., Al[5] or Al[6].

One of the structural differences between C–S–H and C–A–S–H gel is that cross-links are more likely to be formed in the C–A–S–H gel. Cross-links in C–A–S–H are often found in AAS or alkali-activated fly ash paste (calcium poor and aluminum rich) [20, 25]. High curing temperature can also facilitate the formation of cross-links [26]. This structural change, due to its ability of forming a covalent bond between neighboring C–A–S–H layers, may have the potential to significantly improve the mechanical properties of cement-based materials. Hence, it is necessary to further study the nanoscale properties of the cross-linked C–A–S–H. The mechanical properties of the cement hydrate have been studied by many experimental techniques. The commonly used mechanical properties study technique, nanoindentation, can reflect the elastic modulus and hardness of materials at the microscale, making its testing results susceptible to pores of the size of atomic scale in C–A–S–H gel [27, 28]. Recently, Geng et al. [29] have studied the crystal lattice deformations of cross-linked C–A–S–H gels under hydrostatic pressure, utilizing the synchrotron radiation-based high-pressure XRD (HP XRD) technique. This method can intrinsically avoid the influence of the porosity of C–A–S–H gel and the results confirmed that the mechanical strength along c lattice direction increases with the increasing of Al/Si ratio from 0 to 0.1. Furthermore, first-principles calculations from Pellenq [30] on the mechanical properties of the cross-linked [31] and non-cross-linked tobermorite [32] also reveal the enhancement of elastic constant along interlayer direction with the cross-linking of the C–S–H structure.

This paper aims to investigate the mechanical properties and failure mechanisms of the cross-linked C–A–S–H gels along three directions, respectively, using the reactive force field (ReaxFF) molecular dynamics (MD) simulation. Firstly, the cross-linked C–A–S–H models with different Al/Si ratio were built by silicate-aluminate skeleton construction, water adsorption and hydrolytic reaction. Subsequently, the structural evolution in the cross-linked C–A–S–H substrate and hydrogen bond network in interlayer space, as well as the water dynamics, were studied. Finally, the mechanical properties of the C–A–S–H models were obtained by simulating the uniaxial tensile test and the hydrolytic reaction pathways during the tensile process were analyzed to elucidate their failure mechanisms.

6.2 Model Construction

The molecular structure of C–A–S–H gel was constructed following the method proposed in literatures [33, 34]. Merlino's normal 11 Å tobermorite [31] [Ca$_{4.5}$Si$_6$O$_{16}$(OH)·5H$_2$O] was taken as the initial model. Firstly, the 11 Å tobermorite was replicated two times both along the x and y axes to construct a supercell. Then water molecules in this supercell were totally removed and half of the bridging positions in the infinite silicate chain were replaced by aluminate tetrahedra, with hydrogen balancing the negative charge induced by each aluminum substitution for silicon. The charge-balancing hydrogen atom is originally coordinated to bridging oxygen in Al–O–Si bond. Note that the ReaxFF enables protons diffuse to their energetically favored positions and hence this initial configuration would not influence the final equilibrated C–A–S–H structure. Secondly, the aluminosilicate chain in the obtained Al-tobermorite was depolymerized by randomly removing bridging AlO$_2$ to satisfy the Al/Si ratio of 0, 0.05, 0.10, 0.15, and 0.20. Structurally, the integrity of aluminosilicate chains (proportion of bridging sites occupied by SiO$_4$ or AlO$_4$) in these five models are 50%, 62.5%, 75%, 87.5%, and 100%, respectively. Thirdly, the five anhydrous calcium aluminosilicate were relaxed at 0 K by energy minimization using ReaxFF. After the relaxation, water molecules were inserted back into the models by Grand Canonical Monte Carlo (GCMC) method utilizing the CSHFF [35] force field applied by GULP package. By repeatedly inserting, rotating, and deleting molecules at a fixed chemical potential $\mu = 0$ eV and temperature $T = 300$ K, GCMC method can accurately simulate the water adsorption of porous materials at a fictitious reservoir in ambient temperature. Finally, reactive force field molecular dynamics (MD) on the five C–A–S–H models was perform with Reax package [36] in LAMMPS software [37], using the Verlet algorithm to integrate the atomic trajectories, and a time step of 0.25 fs. The first 250 ps of run were employed in isothermal–isobaric (NPT) ensemble at $T = 298$ K and $P = 1$ atom. After the equilibration, we continued further 1000 ps of production dynamics to average the system properties. During this step, the atomic trajectories were recorded every 100 steps for the subsequently structural and dynamics analyses.

After the energy minimization and MD equilibrium simulation, the obtained molecular structures of the five hydrated models are presented in Fig. 6.1. Correspondingly, the atomic density profiles along z-direction of these models are plotted in Fig. 6.2. It can be observed from Fig. 6.1 that the calcium atoms (Cas) and the neighboring oxygen atoms (Os) are assembled to form the Cas–Os layer. On both sides of the Cas–Os layer, defective aluminosilicate chains are grafted. Some bridging sites of the silicate chains of adjacent calcium silicate sheets are cross-linked. With increasing Al/Si ratio of the C–A–S–H model, the number of cross-linked branch structures increases and the polymerization degree of the aluminosilicate chains is also enhanced. When the Al/Si ratio of the C–A–S–H reaches 0.2, all the bridging positions in the aluminosilicate chains are cross-linked. In the cavities among calcium silicate skeleton, interlayer water molecules (Hw, Ow), and calcium atoms (Caw) are randomly distributed.

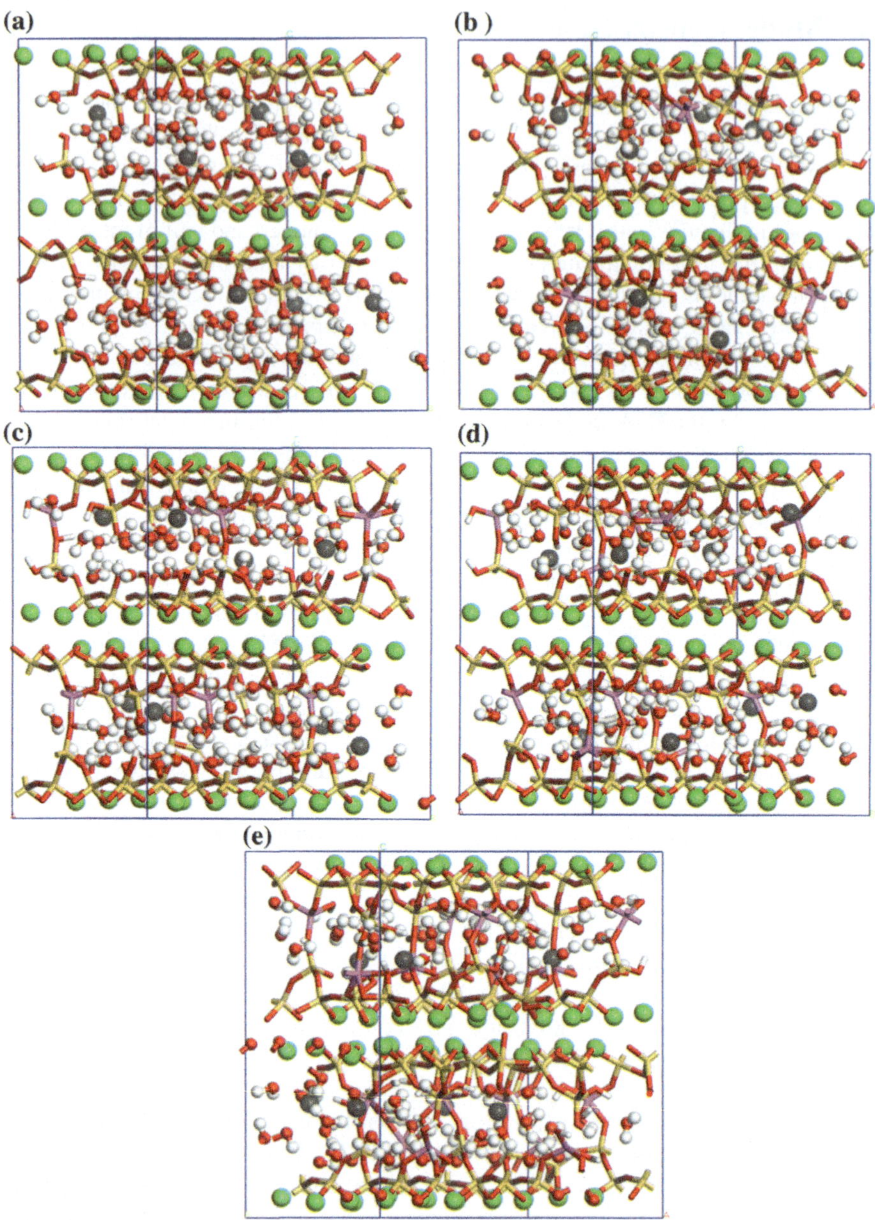

Fig. 6.1 Molecular structure of the C–(A–)S–H gel with Al/Si ratio of **a** 0.00, **b** 0.05, **c** 0.10, **d** 0.15 and **e** 0.20

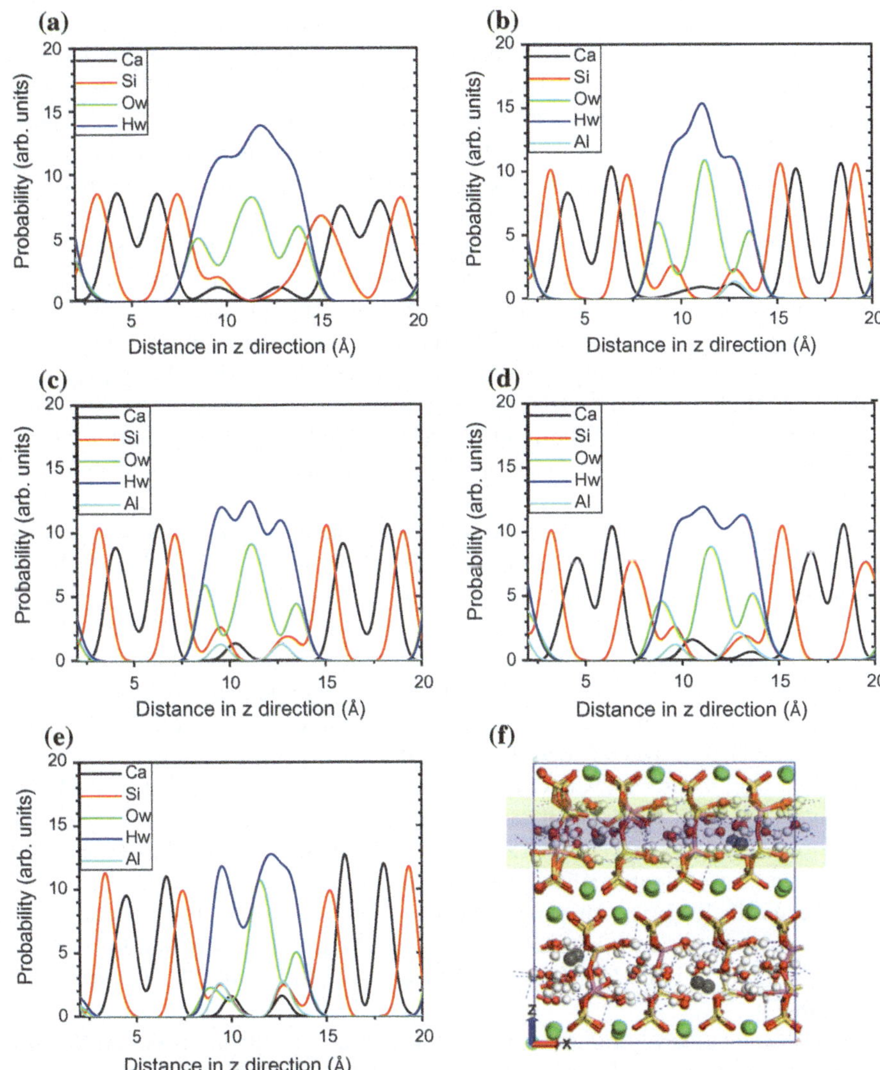

Fig. 6.2 Atomic density profiles of the C–(A–)S–H models with Al/Si ratio of **a** 0.00, **b** 0.05, **c** 0.10, **d** 0.15 and **e** 0.20. **f** Molecular structure of the C–A–S–H gel with Al/Si of 0.15. The three translucent belts indicate triple Ow intensity peaks

The alternative maxima of Ca, Si, Ow, and Hw in the density profiles (Fig. 6.2) suggest the sandwich-like layered structure of C–A–S–H gel. In these profiles, Ow atoms can be distinguished into a trinomial distribution, where the outside two peaks correspond to water molecules adsorbed on C–A–S–H sheets, and the central peak involves water molecules hydrogen bonded with the cross-linked sites, as illustrated in Fig. 6.2f. With respect to Hw atoms, they have broader density peaks than that of Ow atoms. With the increasing of Al/Si ratio, the peaks of aluminum atoms become stronger, implying more aluminum atoms entering the bridging sites in the dreierketten chains. Meanwhile, the density profiles of Ca and Si exhibit sharper peaks that are less overlapped. It indicates that the calcium silicate sheets in the C–A–S–H gel turns more ordered as the structure is cross-linked. It should be noted that the increase of crystallinity with cross-linking of C–A–S–H has also been proven by the XRD studies on synthesized C–A–S–H gel with Al/Si ratio of 0.1 [26]. This is attributed to the fact that the branch structures linking the neighboring layers can enhance the interaction between the layers and maintain the crystal structural integrity.

Interlayer distance and water content within the structure are two important parameters to characterize the structure of C–A–S–H gel. As shown in Fig. 6.3a, the basal spacing of the C–A–S–H models ranges from 12.16 to 11.66 Å, which is in reasonable agreement with experimental results [11, 17, 26]. It can be observed that the cross-linking induced by aluminum incorporation leads to an evident decrease in the basal spacing (<12 Å) of C–A–S–H models, as compared with that of the non-cross-linked C–S–H model (>12 Å). This can be interpreted by the strong structural constraint from covalent bonds between adjacent layers. In the non-cross-linked model, hydrogen bonds and Ca–O ionic bonds bridge the neighboring C–S–H primary layers together, while covalent bonds play predominant role in the interlayer connection of the cross-linked models. The C–A–S–H model with Al/Si of 0.2 in this work has an infinite double silicate chains, which structurally resembles Merlino's 11 Å tobermorite [31]. The interlayer spacing of C–A–S–H model of Al/Si of 0.2 is 11.66 Å, which is higher than that of cross-linked 11 Å tobermorite (11.24 Å). This is due to the longer bond length of Al–O than Si–O (1.85 Å of Al–O > 1.65 Å of Si–O) [38].

The water content evolution of C–A–S–H structure as the function of its Al/Si ratio is shown in Fig. 6.3b. The water amounts within the structure fall in the range of the experimental values obtained by Myers et al. [26] The H_2O/Si ratios of cross-linked C–A–S–H gel is lower than that of C–S–H gel in hydrated cement ($1.3 < H_2O/Si < 1.8$ [39, 40]). In C–S–H gel, the short silicate chains occupy predominated percentage of the silicate morphology. The defective silicate chains result in many vacancy regions in the C–S–H gel where more water molecules can penetrate. This water amount reduction can also be associated with the narrow basal spacing in the cross-linked C–A–S–H gel mentioned above. It is valuable to note that the movement of water molecules in the C–S–H gel pore is responsible for the shrinkage of the cement-based material. The reduction of interlayer water molecules as well as the cross-linking of neighboring layers may effectively resist the creep of the material [41, 42].

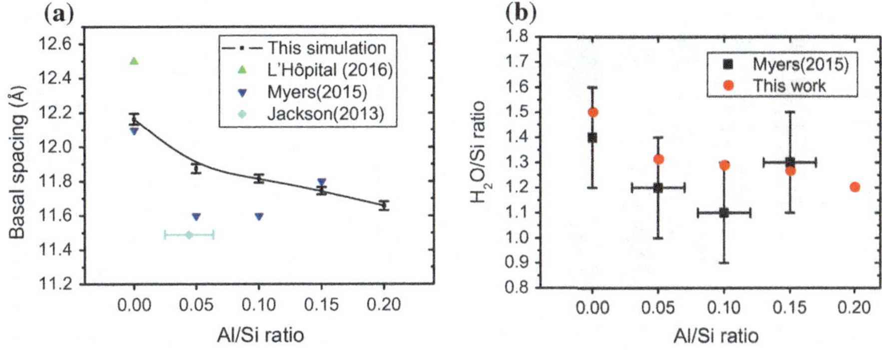

Fig. 6.3 **a** Interlayer distances and **b** water content of C–A–S–H structure at different Al/Si ratios

6.3 Connectivity Factor

The connectivity factor [Qn(mAl)] is used to evaluate the morphology evolution of aluminosilicate chains with increasing Al/Si ratio in the C–A–S–H model. As shown in Fig. 6.4a, Al incorporation causes the change of the Qn distribution in the C–A–S–H. In the Al-free C–S–H model, Q^1 species and Q^2 species account for 40 and 60% of the model, suggesting that silicate tetrahedra are mainly in the form of long silicate chains. With the increase of Al/Si ratio of C–A–S–H model, $Q^2(1Al)$ and $Q^3(1Al)$ species are produced and their proportions gradually increase with the reduction of Q^1 and $Q^2(0Al)$. This clearly reflects that Al incorporation enhances the polymerization degree of silicate tetrahedra. The proportion of $(Q^2 + Q^3 + Q^4)$ as a function of Ca/(Al + Si) ratio is plotted in Fig. 6.4b. Experimental and simulation works from other researchers [16, 26, 28, 43–45] are given as a comparison. It can be observed that the results of our work are in the range of those of others. As shown in Fig. 6.4c, the aluminate species can not only bridge two silicate dimmers to form longer chains, but also cross-links the adjacent silicate chains and produces $Q^3(1Al)$ species. The presence of the Q^3 species transforms the twodimensional layered structure to threedimensional cross-linked structure. The branch structure in our model is different from those in Qomi et al.'s (Fig. 6.4c here vs. Fig. 3d in Ref. [44]). In their model, Al atoms are put into the interlayer region in C–S–H and the Al species connects four neighboring pairing silicate tetrahedra to form the cross-link. The two different branch structures may cause differences in the mechanical properties of C–A–S–H, such as bulk modulus (which is correlated with the basal spacing of the C–A–S–H [46]), the shear deformation behavior (where hinge mechanism plays an important role [30, 47]) and z directional tensile strength (different covalent bond density along the interlayer direction). As shown in Fig. 6.4d, the mean chain length (MCL) of aluminosilicate as a function of Al/Si ratio is computed using the equations described in Ref. [20]. As expected, the aluminosilicate chain length in C–A–S–H increase from 10.00 to infinite, as the Al/Si ratio increases from 0 to 0.20. The calcu-

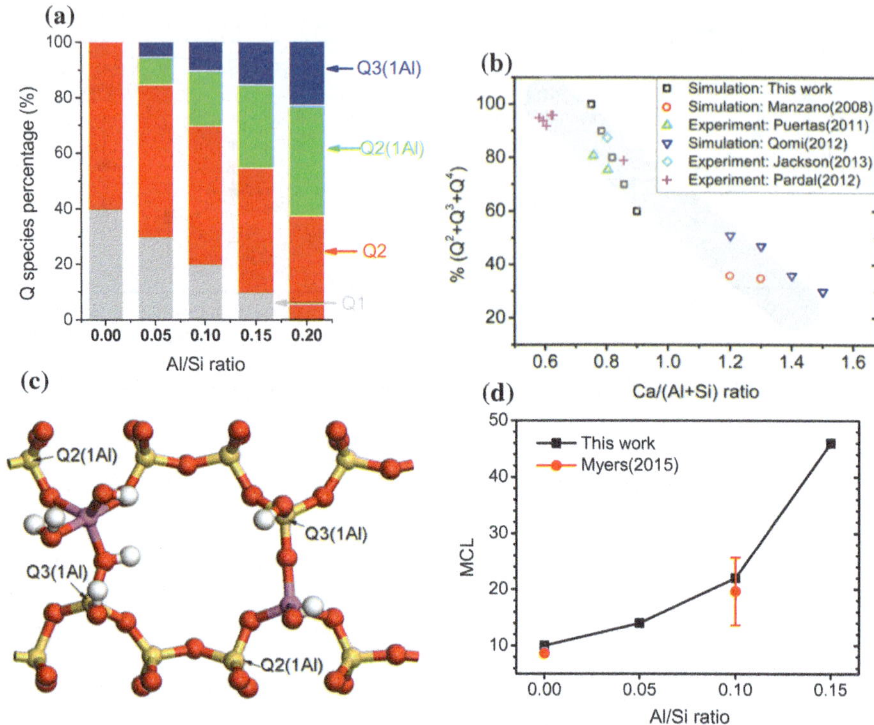

Fig. 6.4 **a** Q species distribution for the C–A–S–H models of different Al/Si ratio; **b** increasing proportion of chain terminals with the increase of Ca/(Al + Si) ratio; **c** molecular structure of the Q species in the aluminosilicate chain. **d** Mean chain length (MCL) evolution of (alumino)silicate chains in the C–(A–)S–H models. Note that the MCL is infinite for C–A–S–H model with Al/Si ratio of 0.2, thereby not denoted in the chart

lated MCL values quantitatively match well with the experimental data [26]. In the experiment, those C–A–S–H samples were cured under 80 °C, and cross-links were found in a sample of the Al/Si ratio of 0.1.

6.4   Coordination Number of Al Atoms

The coordination number (CN) of aluminum atoms in the C–A–S–H models is plotted in Fig. 6.5a. It can be noted that the average CN of aluminum in different models range from 4.5 to 5, including about 3 oxygen atoms in the silicate chains (Os atoms) and less than 2 oxygen atoms from interlayer water molecules or hydroxyl groups. The coordination number of Os atoms is higher in our work than in a previous calculation [48] on non-cross-linked C–A–S–H. It indicates a higher degree of polymerization of aluminosilicate skeleton in the cross-linked C–A–S–H. The four-

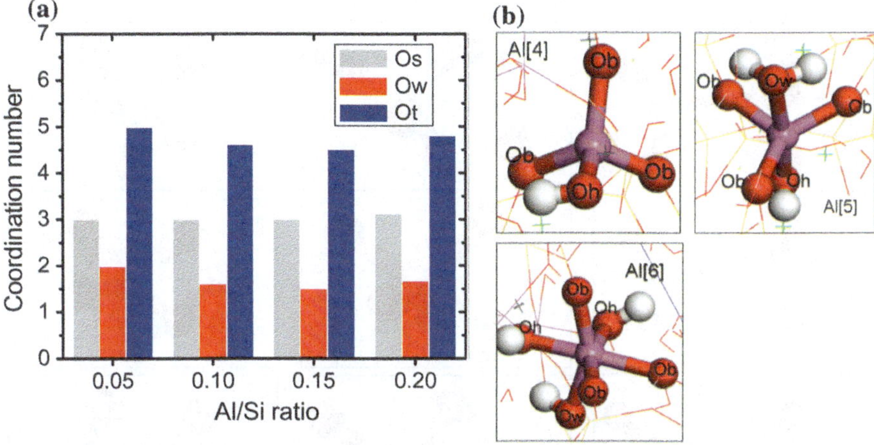

Fig. 6.5 **a** Coordination number of aluminum atom in the C–A–S–H models. Os: oxygen atoms in the silicate chains; Ow: oxygen atoms in the water molecules or hydroxyls. **b** Four-/five-/sixfold coordinated aluminate structures

/five-/sixfold coordinated aluminate structures have been observed in our simulation (see Fig. 6.5b). This consistent with the ^{27}Al NMR analyses of C–A–S–H gel in other experimental studies [13, 14, 16, 17]. Furthermore, in our work, since the CN of Os atoms is almost fixed at 3, the main difference among aluminate species is the CN of Ow atoms, that is, aluminum atoms with higher CN number are coordinated to more Ow atoms. This is in agreement with the NMR test results from Rawal et al. [49] and Andersen et al. [14] that the five-/sixfold coordinated aluminum atoms is mainly attributed to those surrounded by water molecules.

6.5 Structure and Dynamic Properties of Interlayer Water Molecules

Apart from the calcium aluminosilicate skeleton, water molecules confined in the interlayer space are also an important component in the C–A–S–H gel. There are two types of water dissociation reactions in the interlayer regions of C–A–S–H gel. The first one, as depicted in Fig. 6.6a, is the reaction between water and non-bridging oxygen (Onb) atoms in Si–O$^-$ bond. The water molecule diffuses to the silicate chain, and then associates with it by H-bond connection. Subsequently, the water molecule dissociates into H$^+$ and OH$^-$ owing to the strong electronic attraction from Onb atom. While H$^+$ ion bonded to the Onb to form Si–OH group, the remaining OH$^-$ is coordinated to the interlayer Ca^{2+} ion, producing Ca–OH. The second one is the water molecules' association with aluminate tetrahedron. As shown in Fig. 6.6b, the tetra-coordinated aluminum has coordinated one extra water molecule to form penta-

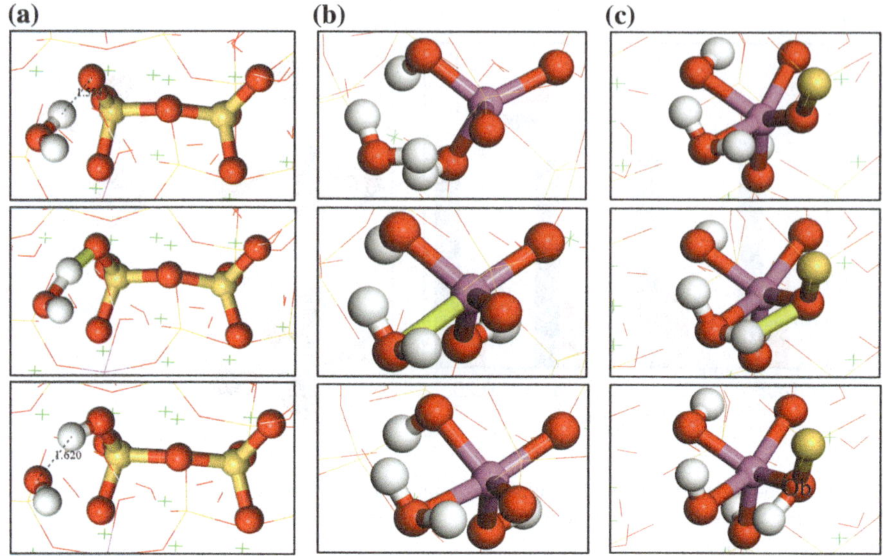

Fig. 6.6 **a** Water dissociation and corresponding Si–OH and Ca–OH formation; **b** water molecule adsorption and coordination variation of aluminum atom; **c** proton transfer from adsorbed water molecule to bridging oxygen atom (Ob)

coordinated aluminum (trigonal bipyramid structure). Furthermore, it is noteworthy that, in some cases, reactions can take place on bridging oxygen (Ob) atom in the Al–O–Si bonds. As shown in Fig. 6.6c, the adsorbed water molecule transfers one of its hydrogen atoms to the Ob atom and then turns to a hydroxyl, while the Ob atom becomes threefold coordinated and an Ob–H hydroxyl is produced. This is quite different from the Ob atom in Si–O–Si bond [33] that has no reactivity with water molecules and shows a hydrophobic nature. The reactivity of Ob in Al–O–Si bond should be attributed to the intermediate ionicity of Al–O bond. Similarly, MD studies on silicate-aluminate glasses also observe the prevalence of oxygen tri-clusters when Al element is incorporated in the glass [50, 51].

The reaction degree, defined as the dissociation ratio of the water molecules, is recorded to estimate the water reactivity in these C–A–S–H models. As shown in Fig. 6.7, the reaction degree increases from 12.5 to 16.7% with increasing Al/Si ratio from 0 to 0.2. It implies that Al incorporation increases the chemical reactivity of water in the C–A–S–H gel. Unexpectedly, the water reactivity enhancement at high polymerization degree is contrary to the fact that water molecules are more likely to dissociate at non-bridging oxygen atoms in the low polymerized silicate-aluminate structure. It should be noted that the incorporation of Al species plays a dual role in the water dissociation. On the one hand, the Al atoms, healing the broken silicate chains, contribute to silicate-aluminate polymerization and reduce the number of non-bridging oxygen sites, which inhibits the first water dissociation reaction near the defective silicate chains. On the other hand, Al substitutes part of

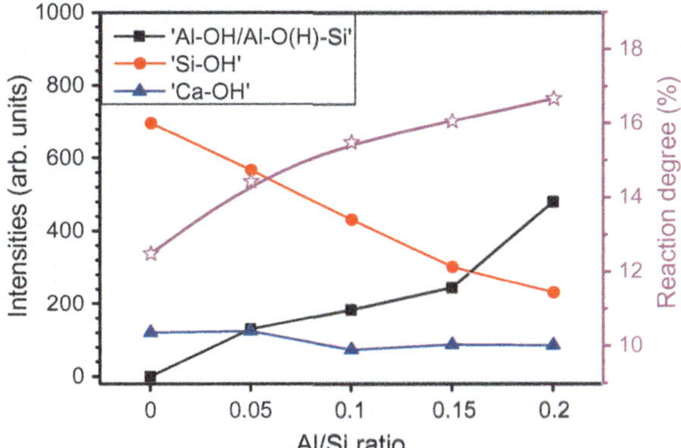

Fig. 6.7 The number of hydroxyl groups and reaction degree evolution with increasing Al/Si ratio

the bridging Si, which destabilizes the Si–O–Si bonds and produce more reactive Ob sites in the Al–O–Si bonds. This contributes to the second water dissociation reaction. The number of hydroxyls evolution is also specified in Fig. 6.7. As shown in the figure, with the increase in Al/Si ratio, while the number of Si–OH groups gradually decreases from 696 to 228, the number of Al–OH/Al–O(H)–Si increases to 480. The increment of Al–OH groups is slightly higher than the decrement of Si–OH. Hence, the increase in reaction degree should be attributed to a more important role that the second hydrolytic reaction plays. At high Al/Si ratio, hydroxyl groups are mainly from the Al–OH/Al–O(H)–Si. It is worth noting that the enhanced water reactivity is not relevant to the increase in polymerization degrees of aluminosilicate chains. If there are the silicon atoms instead of aluminum atoms entering the structure, the polymerization of silicate chains would still increase, but the water reactivity of the structure would not.

The H-bond network in the nanopores of C–A–S–H gel is an important bond that plays a role in linking the neighboring calcium silicate layers [48]. Water molecules ultraconfined in the interlayer region of C–A–S–H are connected to each other or the substrate through hydrogen bonds. To study the H-bond network evolution as a function of increasing Al/Si ratio in C–A–S–H, the average number of H-bonds per water molecule and its five components are calculated. As shown in Fig. 6.8, while the number of Ow–d–Ob (water molecules that donate H-bonds to Ob atoms) gradually increases from 0.40 to 1.00 with increasing Al/Si ratio, the number of Ow–d–Oh decreases from 0.63 to 0.32. This can be related to the progressive increase of Ob species and reduction of Oh species. Furthermore, the number of Ow–a–Oh H-bond rises from 0.45 to 0.59 with increasing Al/Si ratio. This can be explained by the fact that increasing number of Al atoms entering the interlayer structure results in the formation of more Al–OH and Al–O(H)–Si bonds that donates H-bond to neighboring water molecule H-bond. With respect to H-bonds between water molecules, the average number of Ow–a–Ow and Ow–d–Ow bonds simultaneously decrease from 0.65 to 0.41. This bond number reduction is attributed to the branch structure of the connected bridging silicate (aluminate) tetrahedra that block the channel of inter-

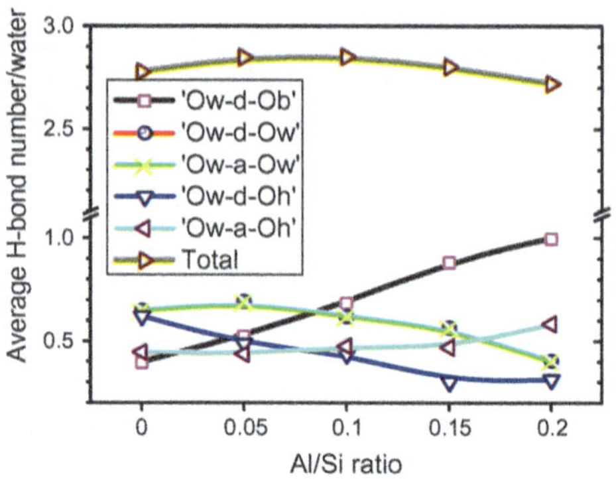

Fig. 6.8 Average H-bond number evolution for different Al/Si ratios ranging from 0 to 0.2

layer water molecules. Overall, with the cross-linking of the aluminosilicate chains in C–A–S–H model, the connections of water molecules are progressively replaced by connections of water and the substrate. This implies that the structural confinement of substrate on the water is strengthened as Al/Si ratio increases. The average H-bond number in various models ranges from 2.72 to 2.86 and does not change much with increasing Al/Si ratio. These values are higher than those obtained in a previous empirical force field (CSHFF) MD simulation [52]. As the CSHFF is unable to characterize the water reaction and proton transfer, none of the silicate tetrahedra are hydroxylated in that work. In our study, however, the hydrolytic reactions producing Si–OH, Al–OH, and Ob–H groups, significantly increase the number of H-bond donating sites.

The RDFs of Ob–Hw for different models are computed to illustrate the bond strength between water molecules and siloxane bonds. As can be seen in Fig. 6.9a, the first peak in the RDF curve for Al-free C–S–H gel is located at 1.92 Å, corresponding to the H-bond between Hw and Ob in Si–O–Si. With increasing Al/Si ratio, the peak located at 1.92 Å does not change much whereas another peak at 1.65 Å appears and turns stronger. The peak with high intensity at 1.65 Å means that the bond distance of Hw and Ob in Al–O–Si (Fig. 6.9c) is shorter than the H-bond length between Hw and Ob in Si–O–Si groups (Fig. 6.9b). This further confirms the stronger hydrophilic nature of Al–O–Si bonds. As a matter of fact, the high reactivity of Ob atoms in Al–O–Si bonds is also an indication of their hydrophilicity.

The interlayer H-bond structure influences the dynamic properties of confined water molecules. The MSD evolution as the function of time is shown in Fig. 6.10 for confined water in five C–A–S–H models. The diffusion coefficients of confined water are computed by linearly fitting the corresponding MSD curve in the diffusive regime, as listed in Table 6.1. The diffusion coefficient of water in C–S–H gel is 7.51×10^{-11} m^2/s, in the same magnitude of the value ($D = 1/60DH_2O = 5.90 \times 10^{-11}$ m^2/s) obtained by proton field cycling relaxometry (PFCR) approach [53] and the value ($D \approx 1.0 \times 10^{-10}$ m^2/s) obtained by quasielastic neutron scattering (QENS) [54]. It means that the interlayer water molecules are strongly restricted by the C–S–H substrates. Due to the restrictions and hydrophilic C–S–H substrate, it is difficult for

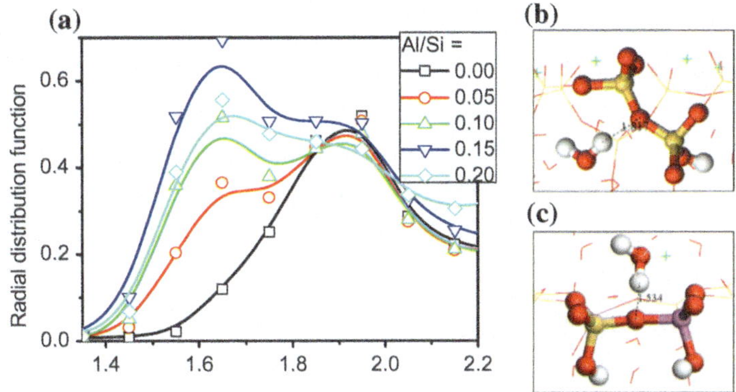

Fig. 6.9 **a** RDFs of Ob–Hw in C–(A–)S–H gels; snapshots of water molecules hydrogen bonded to Ob atom in, **b** Si–O–Si bond (H-bond length of 1.817 Å), and **c** Al–O–Si bond (H-bond length of 1.531 Å)

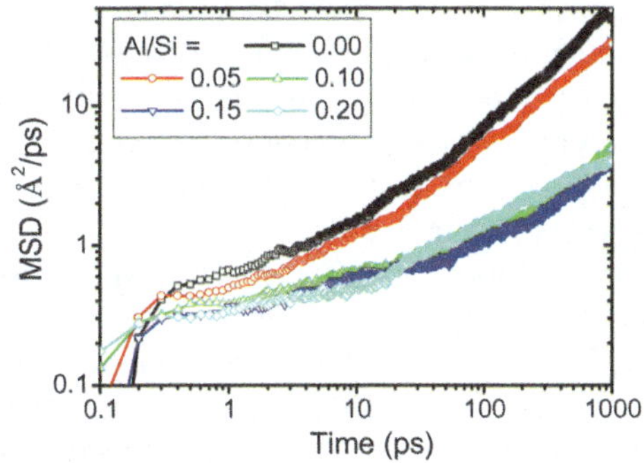

Fig. 6.10   Mean squared displacements (MSD) of interlayer water molecules in different C–A–S–H structures

the confined water to escape from the "cage" constructed by the ionic-covalent bonds and H-bonds. Hence, the mobility of the confined water molecules is significantly reduced as compared with the bulk solution. Furthermore, with the increase of the Al/Si ratio, the diffusion coefficient of water molecules further reduces. In particular, the diffusion coefficient is about only 1/1000 of bulk water value as Al/Si ratio gets 0.2. This reduction in mobility of water molecules is due to the branch cross-linked structure that blocks the connected water transport channel and inhibits the moving of water molecules. With the cross-linking of the adjacent calcium silicate sheets, the accessible volume for water molecules evolves from connected interlayer regions to discontinuous small pores.

Table 6.1 Diffusion coefficients ($\times 10^{-9}$ m²/s) of water molecules in C–A–S–H with different Al/Si ratios

Al/Si ratio	Diffusion coefficients
0	7.51×10^{-2}
0.05	3.62×10^{-2}
0.10	0.72×10^{-2}
0.15	0.57×10^{-2}
0.20	0.32×10^{-2}

6.6   Stress–Strain Relation

In order to study the influence of cross-link on the anisotropic cohesive properties of C–A–S–H, uniaxial tensions along x, y, and z-directions were performed on the five C–A–S–H models. Methods used in a previous mechanical properties investigation were applied [48]. The supercells of the five C–A–S–H were periodically extended along x, y, and z-directions to construct models of the size of about $4 \times 4 \times 4$ nm³. Then 100 ps of MD simulation in NPT ensemble ($T = 298$ K and $P = 1$ atom) was carried out on the models to make the structure achieve equilibrium state, following by gradual elongation along x, y and z-directions, respectively. The strain rates are 0.08/ps in all elongation processes. In addition, configurations were recorded every 100 steps for the analyses of failure mode.

The stress–strain curve, characterizing the constitutive relation, has been applied to investigate the mechanical properties of these C–A–S–H gels, and the results are given in Fig. 6.11. Take the case of C–A–S–H model with Al/Si ratio of 0.1 (Fig. 6.11c), with the progressively increasing of x-directional tensile strain, the stress first rises linearly and then slowly increases. The stress reaches a maximum value of 10.66 GPa at strain of 0.19 Å/Å. After the maximum, the stress drops quickly to 4.83 GPa at strain of 0.27 Å/Å and then the stress declines slowly until the end of the tensile loading. The stress–strain relationship along y-direction is complicated in the post-failure stage. After the first drop, the stress exhibits another increase and reaches a local maximum of 9.67 GPa at strain of 0.36 Å/Å. After the local maximum, the stress–strain curve monotonously decreases. The secondary increase of stress, similar with the strain hardening effect in the tensioned steel material [55], may be related to a structural rearrangement in the calcium silicate sheets [38]. It should be noted that the ladder-like stress decreasing in the post-failure regime for all the five models indicate good plasticity in y-direction for C–A–S–H. On the other hand, as shown in Fig. 6.11c, for the C–A–S–H gel tensioned along z-direction, the stress first increases to 6.53 GPa rapidly as the strain reaches 0.16 Å/Å, and subsequently, the stress continuously decrease to zero at the strain of 0.28 Å/Å. It indicates the brittle nature of the C–A–S–H gel along z-direction.

From the stress–strain curve evolution (Fig. 6.11), it can be observed that the cross-links improve the mechanical properties of the structure along y- and z-directions. In particular, their presence greatly influences the z directional mechanical properties of the C–A–S–H structure. The z directional tensile strength and Young's modulus

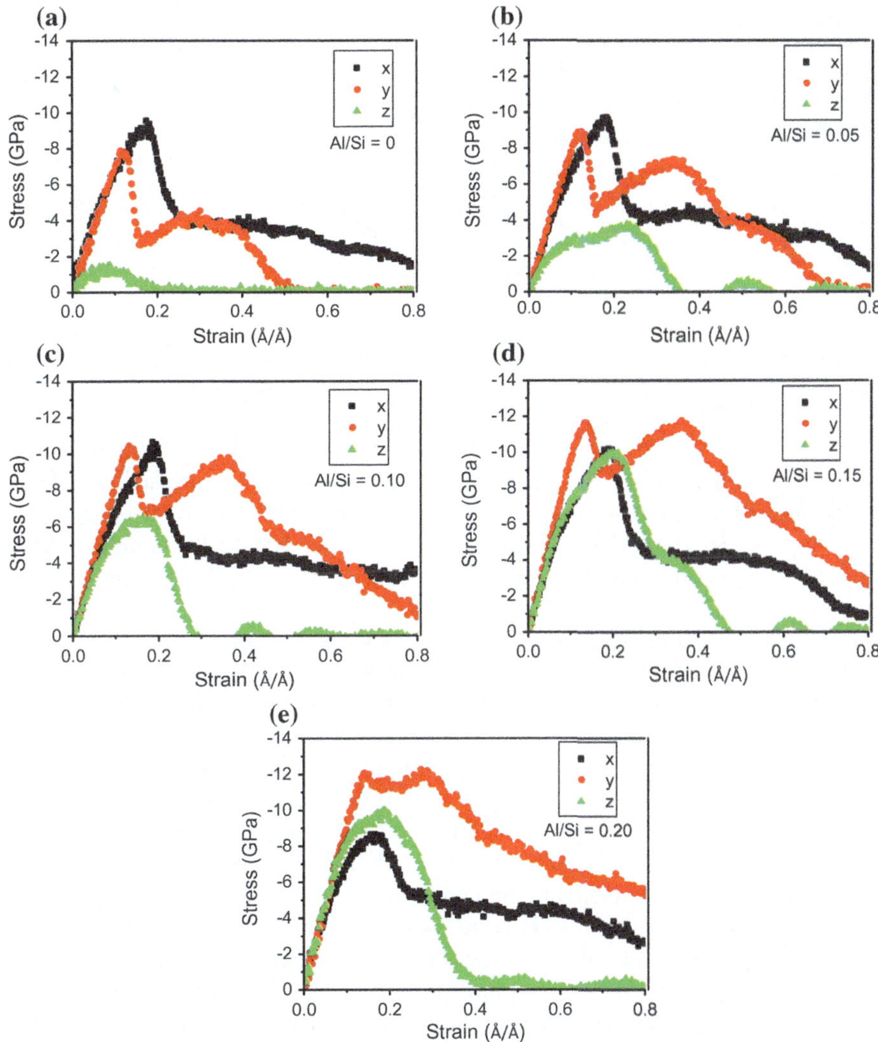

Fig. 6.11 The stress–strain relationships of the C–A–S–H gels with Al/Si ratios of **a** 0.00, **b** 0.05, **c** 0.10, **d** 0.15, and **e** 0.20

are calculated from the stress–strain curves of the models and plotted in Fig. 6.12. As shown in Fig. 6.12, with the increases of Al/Si ratio and cross-linking degree, the tensile strength and the Young's modulus of C–A–S–H along z-direction increase by 296%, and 550%, respectively. It can be concluded that the interlayer strength of the C–A–S–H is significantly enhanced with the cross-linking of the aluminosilicate chains. This result is in agreement with the previous high-pressure XRD technique experimental observations [29], where the lattice parameters evolution of cross-

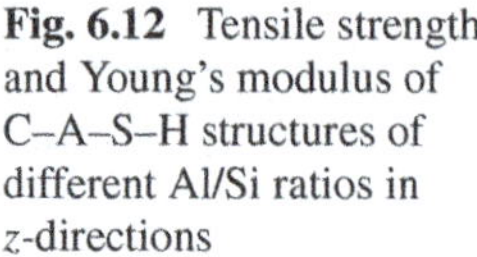

Fig. 6.12 Tensile strength and Young's modulus of C–A–S–H structures of different Al/Si ratios in z-directions

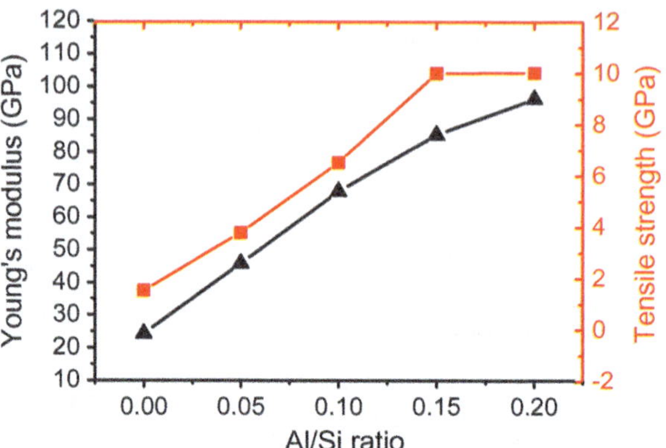

linked C–A–S–H gel under increasing hydrostatic pressure was investigated. Their results show that the compressibility of c lattice parameter of C–A–S–H decreases from 7 to 11% per 10 GPa for Al-free sample to 4% per 10 GPa for sample of Al/Si ratio of 0.1, indicating that Al-induced cross-links stiffen the c axis strength of the sample.

The improvement of mechanical properties of C–A–S–H gel can be explained by polymerization of aluminosilicate chains at high Al/Si ratio. As well known, the calcium silicate sheets in C–S–H gel play the role of a backbone. For the Al-free C–S–H model, silicate tetrahedra are in the form of dimers or pentamers, and the interaction between adjacent layers in C–S–H is ionic or hydrogen bonds between zeolitic content (water and Ca^{2+}) and primary layer. Due to the presence of a large number of H-bonds, the material is soft in z-direction. With the bridging sites gradually occupied by Al atoms, the Al–O and Si–O covalent bonds with high bond strength [30, 56] play a critical role in bridging the neighboring calcium silicate layers together. Additionally, the improvement in H-bond stability at high Al/Si ratio, as discussed in dynamic properties of interlayer water, can also contribute to the increasing strength for C–A–S–H. Bonnaud et al. [57] have proposed that the repulsive force of the interlayer water molecules in the C–S–H gel contributes to the fluid pressure that tends to disjoin the C–S–H principal layers. The frequent formation and breakage of H-bonds destabilize calcium silicate skeleton to some extent. As discussed in previous section, the cross-links between neighboring layers decrease mobility of the interlayer water molecules and hence enhance the stability of H-bond network. This can also contribute to the increase in z directional mechanical strength of C–A–S–H gel.

6.7 Deformation of the Structure

To better understand the failure process of C–A–S–H gel, the deformed molecular structures are shown in Fig. 6.13 for the C–A–S–H model with Al/Si ratios of 0.0 and 0.2 from the strain of 0.0 to 0.3 Å/Å during tension along z-direction. As shown in Fig. 6.13a, the non-cross-linked C–S–H model is easy to stretch fracture. The crack is initiated and developed rapidly in the interlayer region and no pronounced deformation is observed in the calcium silicate sheets during the elongation process. This means that the interactions between calcium silicate layers and the zeolitic content (interlayer water molecules and Ca^{2+} ions) are weak, which makes non-cross-linked C–S–H susceptible to the z directional loading. With respect to the cross-linked C–A–S–H model, the failure process of the structure follows another pattern, as shown in Fig. 6.13b. The Al–O and Si–O bonds in the cross-linked sites are extended and stretched open to take the external loading when the C–A–S–H model is elongated within the strain range of 0.0 and 0.1 Å/Å. At the strain of 0.2 Å/Å, some of the cross-linked sites are broken and small defects in aluminosilicate chains are initiated. It also can be observed that the ordered calcium silicate layers are slightly disturbed due to the external loading. Note that although the Al–O bond is stronger than the Ca–O bond, the calcium silicate main layer with higher bond density still exhibits a mechanical strength superior to the cross-link sites. Thus the fracture of the C–A–S–H structure prefers to occur at the interlayer cross-links. With the further increase in the strain, the cracks grow and coalesce through the defective cross-linked sites. When the strain reaches 0.3 Å/Å, the small cracks connect together to form a large crack, and C–A–S–H structure is finally fractured. The elongation process results in large deformation in calcium silicate layers, and the transformation from a layered crystal phase to an amorphous one. This indicates that, different from C–S–H, the branch structure in aluminosilicate chains can transfer the tensile loading to the calcium silicate substrate, and both the interlayer region and the neighboring calcium silicate sheets can resist the tensile loading together. Hence, the cross-linked C–A–S–H model exhibits a stronger mechanical behavior than that of non-cross-linked C–S–H.

To better understand the failure mechanism of the C–A–S–H under tensile loading, the variation of Q^n species is calculated to give a description on the structural evolution of the aluminosilicate skeleton during the tensile process. As shown in Fig. 6.14a, for the non-cross-linked C–S–H gel, since the cohesive force between the layers mainly comes from the hydrogen and ionic bonds, the tensile process seems not to exert any influence on the polymerization of silicate tetrahedra. The proportions of Q species are constant during the elongation (Fig. 6.14a). The Q species percentage change during the tension process along z-direction in the cross-linked models. Take the C–A–S–H with Al/Si ratio of 0.1 for an example. As shown in Fig. 6.14c, when the strain ranges from 0 to 0.10 Å/Å, the percentages of Q species remains unchanged. This stage corresponds to the stress linear increase stage in stress–strain curve (Fig. 6.11c). In this stage, the Al–O and Si–O bonds in the cross-links are elongated and Al–O–Si angles are stretched open to carry the loading. When the strain

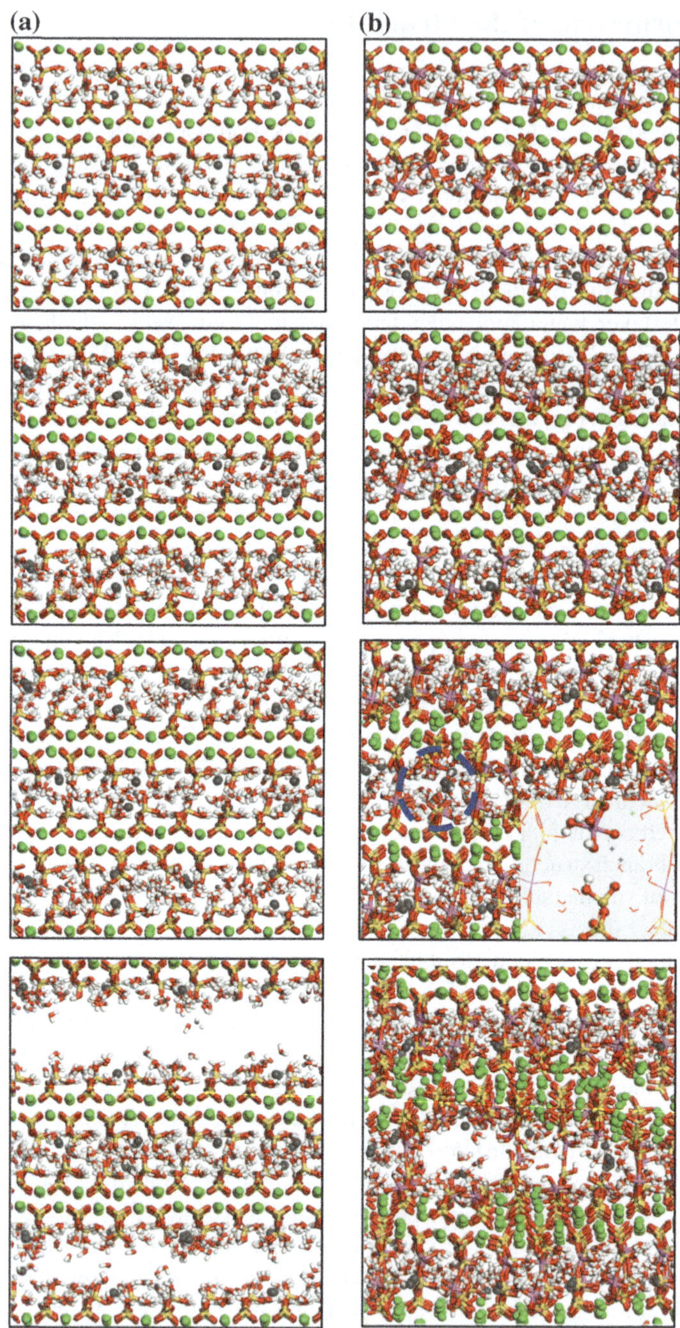

Fig. 6.13 Molecular structure evolution (view along y-direction) of C–A–S–H gel with Al/Si ratio of **a** 0 and **b** 0.2 during elongation along z-direction. From top to bottom: the C–A–S–H gel at strain of 0.0, 0.1, 0.2, and 0.3 Å/Å

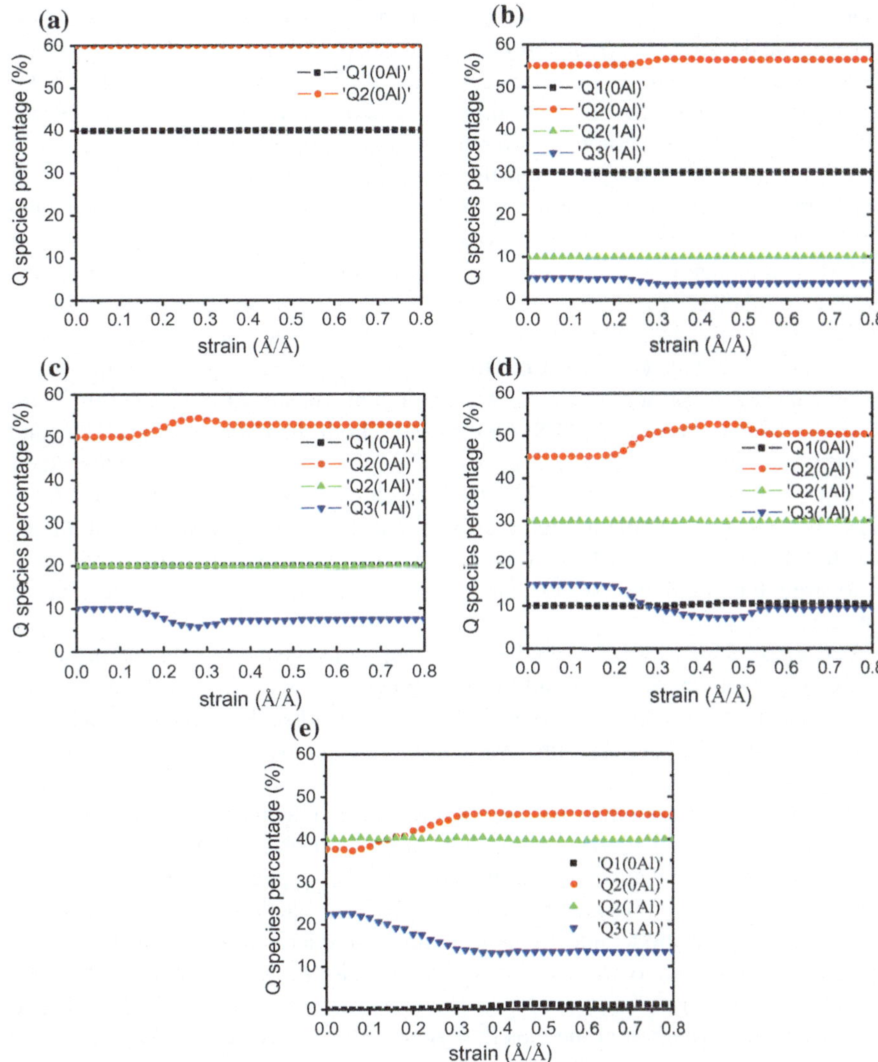

Fig. 6.14 Q species evolution of C–A–S–H models under z-direction elongation with Al/Si ratio of **a** 0.00, **b** 0.05, **c** 0.10, **d** 0.15, and **e** 0.20

is larger than 0.10 Å/Å, the Q^3 species begins to decrease and Q^2 increases. This suggests that the Al–O bonds are broken, and the depolymerization of the aluminate-silicate branch structure results in the breakage of cross-links in the model. The Q^2 species percentage reaches the maximum value at the strain of 0.28 Å/Å, where the C–A–S–H structure is a complete stretched fracture. As shown in Fig. 6.14c, some Q^2 species turns back to Q^3 within the strain ranging from 0.28 to 0.35 Å/Å. It means that after the fracture of the C–A–S–H, there is a structural rearrangement of the calcium aluminosilicate skeleton at each newly formed fracture surface. The $Q^2(1Al)$ percentage is almost constant during the elongation process, implying the depolymerization of branch structure is occurred in the cross-link positions, while the aluminosilicate chains on both sides of the cracks remain their long chain structures. Furthermore, it can be observed that the change of Q^n species is more pronounced with increasing Al/Si ratio of the C–A–S–H model. As shown in Fig. 6.14b, e, for the C–A–S–H structure with Al/Si ratio of 0.05, both the reduction in $Q^3(1A)$ and the increment in $Q^2(0Al)$ are 1.25%, while the reduction in $Q^3(1A)$ and the increment in $Q^2(0Al)$ are 9.00% and 8.02% for C–A–S–H model with Al/Si ratio of 0.20, respectively. It implies that at higher cross-linking degree, more silicate-aluminate branch structures in the C–A–S–H model carry the tensile loading and depolymerize to strengthen the interlayer mechanical properties and retard the fracture in the material.

As suggested in the previous study [58], water molecules' intrusion into the structures can "attack" the Si–O–Si and Si-O-Ca bonds, and weaken their loading resistance. This process involves the dissociation of water molecules and the formation of hydroxyl groups, thus these changes were also recorded to investigate the hydrolytic reaction during tensile loading.

In addition to silicate-aluminate depolymerization reaction, the hydrolytic reaction happens in the C–A–S–H gel during the tensile process. As shown in Fig. 6.15, the change of hydroxyls and H_2O number is recorded as a function of the tensile strain along z-direction. As shown in Fig. 6.15a, no hydrolytic reactions are observed in non-cross-linked C–S–H model under tensile loading. In order to illustrate the hydrolytic reactions during failure of the cross-linked models, the C–A–S–H model with Al/Si ratio of 0.10 is taken as the example. As shown in Fig. 6.15c, there is no Si–OH and Ca–OH groups formation at the elastic regime of the stress–strain relation. On the other hand, the number of water molecules is reduced at this stage, and the number of Al–OH and Al–O(H)–Si groups increases. It means that water dissociates to produce the aluminate hydroxyl groups. This is attributed to the local structural change for aluminate polyhedra and protonation of bridging oxygen atoms. As discussed above, the aluminum atom can increase its coordinate number by associating with extra hydroxyl groups and the Ob atom in Al–O–Si bonds is able to react with proton to form the Al–O(H)–Si tri-cluster (see Fig. 6.6c). This process involves dissociation of one water molecule and the formation of one Al–OH group and one Al–O(H)–Si group. Thermodynamically, the tensile loading can reduce the energy barrier that is needed to activate hydrolytic reaction of confined water molecules [59]. Therefore, when the Al–O bonds are elongated and Al–O–Si angles stretched open, this disso-

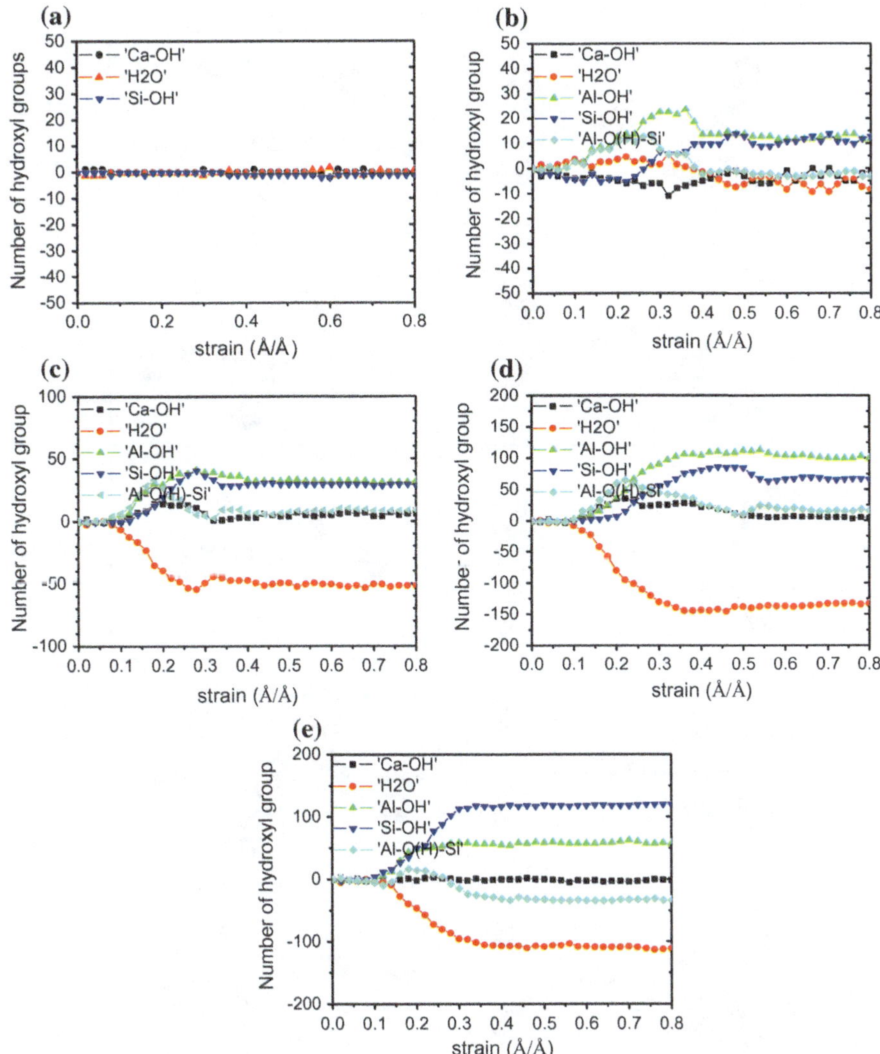

Fig. 6.15 Number of Ca–OH, Si–OH, Al–OH, and Al–O(H)–Si and H_2O evolution of C–A–S–H models under z-direction elongation with Al/Si ratio of **a** 0.00, **b** 0.05, **c** 0.10, **d** 0.15, and **e** 0.20

ciation reaction is promoted. When the strain reaches 0.10 Å/Å, the aluminosilicate chains start to break and the hydrolytic reactions are widely happening.

There are two reaction pathways for the hydrolytic reaction of Al–O–Si bonds, depending on the local structure of cross-links. The reaction pathways can be illustrated in Fig. 6.16a, b. In the first pathway (Pathway 1), the water molecule adsorbed on the aluminate species transfers one proton to Ob atom in the Al–O–Si bond to form an oxygen tri-cluster. The oxygen tri-cluster weakens the strength of Al–O

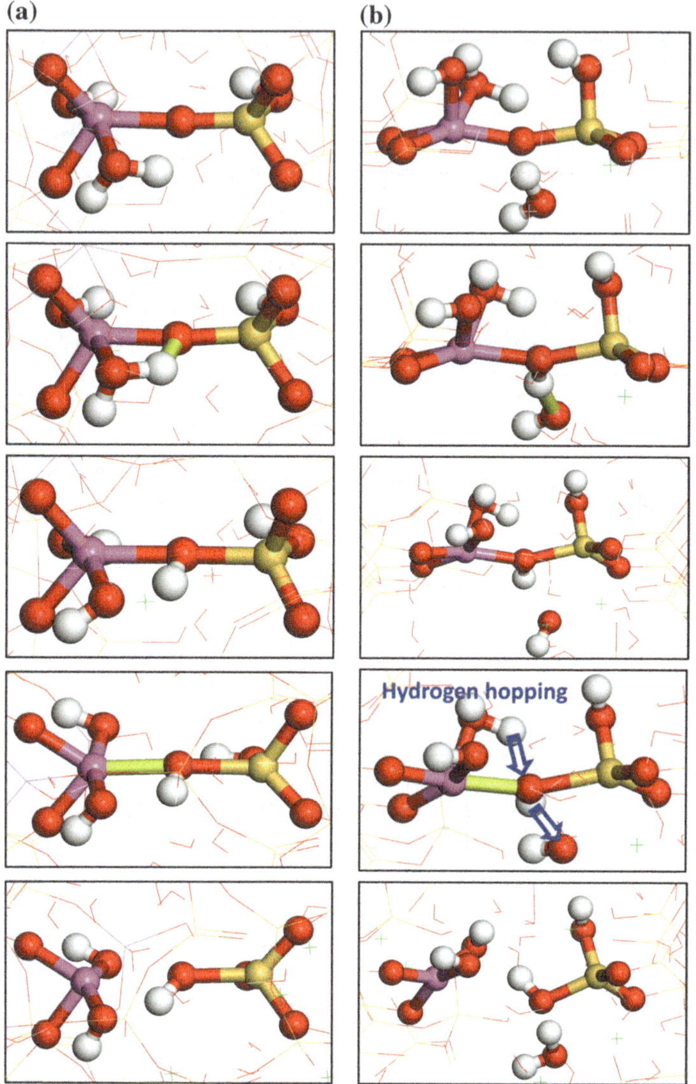

Fig. 6.16 Two hydrolytic reaction pathways of Al–O–Si bonds: **a** pathway 1; **b** pathway 2

bond, and then the Al–Ob–Si bond is broken. As a consequence, Si–OH group is produced after the silicate-aluminate chain breakage. The intermediate products in this reaction pathway are Al–OH and Al–O(H)–Si groups and the reaction products are Al–OH and Si–OH groups. In the second pathway (Pathway 2), it is the neighboring interlayer water molecule that gives one of its hydrogen atoms to Ob atom to form the tri-cluster. Subsequently, the cross-link Al–O bond is stretched broken. Then there is one hydrogen atom "hoping" from the adsorbed water molecule to the free hydroxyl, resembling the process described in Ref. [60]. In this reaction pathway, while the intermediates are Ca–OH and Al–O(H)–Si groups, the reaction products are Si–OH and Al–OH. It should be noted that the hydrolytic reaction of Al–O–Si bond is a two-stage reaction including water adsorption stage and bond breakage stage, where intermediate products can exist for a long time. As shown in Fig. 6.16, both the two reactions mentioned above produce Al–OH and Si–OH groups. In both reactions, the water molecules attack the tensioned Si–O–Al bonds and accelerate the separation between neighboring Q^3 and aluminate species. In this respect, based on the reaction mechanism analysis, the water dissociation reaction and the cross-linking species depolymerization reaction are interplayed and enhance the reaction degree between each other.

Furthermore, the number of Si–OH groups formation during elongation increases as the Al/Si ratio of C–A–S–H increases, illustrating that more cross-links are broken during the failure process of the cross-linked C–A–S–H model with higher Al/Si ratios. This matches well with the Q species percentages evolution, implying that the C–A–S–H model can be strengthened along z-direction by the cross-links.

6.8 Chapter Summary

By utilizing the reactive force field molecular dynamics simulation, the structures, dynamics, mechanical properties, and failure mechanism of cross-linked C–A–S–H gel have been studied. Several conclusions can be made as follows:

(1) The incorporation of aluminate species in the C–A–S–H gel modifies the silicate-aluminate skeleton and interlayer water molecules. On the one hand, Al–O bond can polymerize with defective silicate chains, transforming the layered structure to the cross-linking branch structure. On the other hand, the bridging oxygen atoms in Al–O–Si bonds react with hydrogen atoms to form hydroxyl of Al–OH and Al–O(H)–Si. With increasing Al/Si ratio, more water molecules are dissociated and associate with Al atoms, producing Al–OH and Al–O(H)–Si.

(2) Al–O–Si cross-links can not only exert geometry restriction on the interlayer water molecules and block the connectivity of the nanometer channel, but also increases the substrate affinity toward water molecules. It results in a dramatically decreasing mobility of the interlayer water molecules in C–A–S–H models with higher Al/Si ratios.

(3) Based on the uniaxial tensile testing, due to the branch structure in the interlayer region, both the tensile strength and Young's modulus of the C–A–S–H gel along interlayer direction are significantly improved with increasing Al/Si ratio. The H-bond and the ionic bond between the interlayer regions are replaced by the Al–O and Si–O covalent bonds, which strengthen the connections of aluminosilicate skeleton.

(4) During the failure process of the structure, the external loading combined with attacks from water molecules is the main reasons for the bond breaking. The

hydrolytic reaction of Al–O–Si bond is a two-stage reaction. Firstly, the Ob atom in Al–O–Si bond will "borrow" proton from either the adsorbed water molecule or the free interlayer water molecules to form an Ob–H hydroxyl. Secondly, when the external loading reaches a certain level, the Al–O bond in Al–O–Si cross-link sites will be broken. Finally, the hydrolytic reaction leads to one Al–OH and Si–OH formation and one water molecule dissociation.

References

1. Gartner, E., & Hirao, H. (2015). A review of alternative approaches to the reduction of CO_2 emissions associated with the manufacture of the binder phase in concrete. *Cement and Concrete Research, 78,* 126–142.
2. Gao, Y., Hu, C., Zhang, Y., Li, Z., & Pan, J. (2018). Investigation on microstructure and microstructural elastic properties of mortar incorporating fly ash. *Cement & Concrete Composites, 86.*
3. Hu, C., Li, Z., Gao, Y., Han, Y., & Zhang, Y. (2014). Investigation on microstructures of cementitious composites incorporating slag. *Advances in Cement Research, 26*(26), 222–232.
4. Hu, C. (2014). Microstructure and mechanical properties of fly ash blended cement pastes. *Construction and Building Materials, 73,* 618–625.
5. Ogawa, S., Nozaki, T., Yamada, K., Hirao, H., & Hooton, R. D. (2012). Improvement on sulfate resistance of blended cement with high alumina slag. *Cement and Concrete Research, 42*(2), 244–251.
6. Juenger, M. C. G., & Siddique, R. (2015). Recent advances in understanding the role of supplementary cementitious materials in concrete. *Cement and Concrete Research, 78,* 71–80.
7. Provis, J. L., & Bernal, S. A. (2014). Geopolymers and related alkali-activated materials. *Annual Review of Materials Research, 44*(44), 299–327.
8. Shi, C., Roy, D., & Krivenko, P. (2005). *Alkali-activated cements and concretes.*
9. Lothenbach, B., Scrivener, K., & Hooton, R. D. (2011). Supplementary cementitious materials. *Cement and Concrete Research, 41*(12), 1244–1256.
10. Chen, W., & Brouwers, H. J. H. (2007). The hydration of slag. Part 1: Reaction models for alkali-activated slag. *Journal of Materials Science, 42*(2), 428–443.
11. Jackson, M. D., Chae, S. R., Mulcahy, S. R., Meral, C., Taylor, R., Li, P., et al. (2012). Unlocking the secrets of Al-tobermorite in Roman seawater concrete. *American Mineralogist, 98*(10), 1669–1687.
12. Faucon, P., Delagrave, A., Petit, J. C., Richet, C., Marchand, J. M., & Zanni, H. (1999). Aluminum incorporation in calcium silicate hydrates (C–S–H) depending on their Ca/Si ratio. *Journal of Physical Chemistry B, 103*(37).
13. Sun, G. K., Young, J. F., & Kirkpatrick, R. J. (2006). The role of Al in C–S–H: NMR, XRD, and compositional results for precipitated samples. *Cement and Concrete Research, 36*(1), 18–29.
14. Andersen, M. D., Jakobsen, H. J., & Skibsted, J. (2006). A new aluminium-hydrate species in hydrated Portland cements characterized by Al and Si MAS NMR spectroscopy. *Cement and Concrete Research, 36*(1), 3–17.
15. Manzano, H., Dolado, J. S., & Ayuela, A. (2009). Aluminum incorporation to dreierketten silicate chains. *Journal of Physical Chemistry B, 113*(9), 2832.
16. Pardal, X., Brunet, F., Charpentier, T., Pochard, I., & Nonat, A. (2012). ^{27}Al and ^{29}Si solid-state NMR characterization of calcium–aluminosilicate–hydrate. *Inorganic Chemistry, 51*(3), 1827.
17. L'Hôpital, E., Lothenbach, B., Saout, G. L., Kulik, D., & Scrivener, K. (2015). Incorporation of aluminium in calcium–silicate–hydrates. *Cement and Concrete Research, 75,* 91–103.
18. Churakov, S. V., & Labbez, C. (2017). Thermodynamics and molecular mechanism of Al incorporation in calcium silicate hydrates. *Journal of Physical Chemistry C, 121*(8).

19. Richardson, I. G. (2004). Tobermorite/jennite- and tobermorite/calcium hydroxide-based models for the structure of C–S–H: Applicability to hardened pastes of tricalcium silicate, β-dicalcium silicate, Portland cement, and blends of Portland cement with blast-furnace slag, metakaol. *Cement and Concrete Research, 34*(9), 1733–1777.
20. Myers, R. J., Bernal, S. A., Nicolas, R. S., & Provis, J. L. (2013). Generalized structural description of calcium-sodium aluminosilicate hydrate gels: The cross-linked substituted tobermorite model. *Langmuir, 29*(17), 5294–5306.
21. Taylor, H. F. W. (1986). Proposed structure for calcium silicate hydrate gel. *Journal of the American Ceramic Society, 69*(6), 464–467.
22. Taylor, H. F. W. (1956). Relationships between calcium silicates and clay minerals. *Clay Minerals, 3*(16), 98–111.
23. Richardson, I. G., Brough, A. R., Brydson, R., Groves, G. W., & Dobson, C. M. (1993). Location of aluminum in substituted calcium silicate hydrate (C–S–H) gels as determined by ^{29}Si and ^{27}Al NMR and EELS. *Journal of the American Ceramic Society, 76*(9), 2285–2288.
24. Pegado, L., Labbez, C., & Churakov, S. V. (2014). Mechanism of aluminium incorporation into C–S–H from ab initio calculations. *Journal of Materials Chemistry A, 2*(10), 3477–3483.
25. Bernal, S. A., Provis, J. L., Walkley, B., Nicolas, R. S., Gehman, J. D., Brice, D. G., et al. (2013). Gel nanostructure in alkali-activated binders based on slag and fly ash, and effects of accelerated carbonation. *Cement and Concrete Research, 53*(2), 127–144.
26. Myers, R. J., L'Hôpital, E., Provis, J. L., & Lothenbach, B. (2015). Effect of temperature and aluminium on calcium (alumino)silicate hydrate chemistry under equilibrium conditions. *Cement and Concrete Research, 68*, 83–93.
27. Ulm, F. J., Vandamme, M., Bobko, C., Ortega, J. A., Tai, K., & Ortiz, C. (2007). Statistical indentation techniques for hydrated nanocomposites: Concrete, bone, and shale. *Journal of the American Ceramic Society, 90*(9), 2677–2692.
28. Puertas, F., Palacios, M., Manzano, H., Dolado, J. S., Rico, A., & Rodríguez, J. (2011). A model for the C–A–S–H gel formed in alkali-activated slag cements. *Journal of the European Ceramic Society, 31*(12), 2043–2056.
29. Geng, G., Myers, R. J., Li, J., Maboudian, R., Carraro, C., Shapiro, D. A., et al. (2017). Aluminum-induced dreierketten chain cross-links increase the mechanical properties of nanocrystalline calcium aluminosilicate hydrate. *Scientific Reports, 7*.
30. Shahsavari, R., Buehler, M. J., Pellenq, J. M., & Ulm, F. J. (2009). First-principles study of elastic constants and interlayer interactions of complex hydrated oxides: Case study of tobermorite and jennite. *Journal of the American Ceramic Society, 92*(10), 2323–2330.
31. Merlino, S., Bonaccorsi, E., & Armbruster, T. (2001). The real structure of tobermorite 11Å: Normal and anomalous forms, OD character and polytypic modifications. *European Journal of Mineralogy, 13*(3), 65–80.
32. Hamid, S. A. (1981). The crystal structure of the 11Å natural tobermorite $Ca_{2.25}[Si_3O_{7.5}(OH)_{1.5}]\cdot1H_2O$. *Zeitschrift für Kristallographie—Crystalline Materials, 154*(1–4), 189–198.
33. Manzano, H., Moeini, S., Marinelli, F., Duin, A. C. T. V., Ulm, F. J., & Pellenq, J. M. (2012). Confined water dissociation in microporous defective silicates: Mechanism, dipole distribution, and impact on substrate properties. *Journal of the American Chemical Society, 134*(4), 2208–2215.
34. Pellenq, R. J., Kushima, A., Shahsavari, R., Van Vliet, K. J., Buehler, M. J., Yip, S., et al. (2009). A realistic molecular model of cement hydrates. *Proceedings of the National Academy of Sciences of the United States of America, 106*(38), 16102.
35. Shahsavari, R., Pellenq, J. M., & Ulm, F. J. (2011). Empirical force fields for complex hydrated calcio-silicate layered materials. *Physical Chemistry Chemical Physics, 13*(3), 1002–1011.
36. Aktulga, H. M., Fogarty, J. C., Pandit, S. A., & Grama, A. Y. (2012). Parallel reactive molecular dynamics: Numerical methods and algorithmic techniques. *Parallel Computing, 38*(4–5), 245–259.
37. Plimpton, S. (1995). Fast parallel algorithms for short-range molecular dynamics. *Journal of Computational Physics, 117*(1), 1–19.

38. Hou, D., Zhao, T., Ma, H., & Li, Z. (2015). Reactive molecular simulation on water confined in the nanopores of the calcium silicate hydrate gel: Structure, reactivity, and mechanical properties. *Journal of Physical Chemistry C, 119*(3), 1346–1358.
39. Jennings, H. M. (2008). Refinements to colloid model of C–S–H in cement: CM-II. *Cement and Concrete Research, 38*(3), 275–289.
40. Muller, A. C. A., Scrivener, K. L., Gajewicz, A. M., & Mcdonald, P. J. (2013). Densification of C–S–H measured by ^{1}H NMR relaxometry. *Journal of Physical Chemistry C, 117*(1), 403–412.
41. Ulm, F. J., Bazant, Z. P., & Wittmann, F. H. (2001). Micromechanical analysis of creep and shrinkage mechanisms. In *Creep, shrinkage and durability of concrete and other quasi-brittle materials* (p. 13). Oxford: Elsevier.
42. Pellenq, J. M., Lequeux, N., & Damme, H. V. (2008). Engineering the bonding scheme in C–S–H: The iono-covalent framework. *Cement and Concrete Research, 38*(2), 159–174.
43. Manzano, H., Dolado, J. S., Griebel, M., & Hamaekers, J. (2010). A molecular dynamics study of the aluminosilicate chains structure in Al-rich calcium silicate hydrated (C–S–H) gels. *Physica Status Solidi, 205*(6), 1324–1329.
44. Qomi, M. J. A., Ulm, F. J., & Pellenq, J. M. (2012). Evidence on the dual nature of aluminum in the calcium–silicate–hydrates based on atomistic simulations. *Journal of the American Ceramic Society, 95*(3), 1128–1137.
45. Jackson, M. D., Moon, J., Gotti, E., Taylor, R., Chae, S. R., Kunz, M., et al. (2013). Material and elastic properties of Al-tobermorite in ancient roman seawater concrete. *Journal of the American Ceramic Society, 96*(8), 2598–2606.
46. Geng, G., Myers, R. J., Qomi, M., & Monteiro, P. (2017). Densification of the interlayer spacing governs the nanomechanical properties of calcium–silicate–hydrate. *Scientific Reports, 7*(1).
47. Oh, J. E., Clark, S. M., & Monteiro, P. J. M. (2011). Does the Al substitution in C–S–H(I) change its mechanical property? *Cement and Concrete Research, 41*(1), 102–106.
48. Hou, D., Hu, C., & Li, Z. (2017). Molecular simulation of the ions ultraconfined in the nanometer-channel of calcium silicate hydrate: Hydration mechanism, dynamic properties, and influence on the cohesive strength. *Inorganic Chemistry, 56*(4).
49. Rawal, A., Smith, B. J., Athens, G. L., Edwards, C. L., Roberts, L., Gupta, V., et al. (2010). Molecular silicate and aluminate species in anhydrous and hydrated cements. *Journal of the American Chemical Society, 132*(21), 7321–7337.
50. Benoit, M., Ispas, S., & Tuckerman, M. E. (2001). Structural properties of molten silicates from ab initio molecular-dynamics simulations: Comparison between $CaO–Al_2O_3–SiO_2$ and SiO_2. *Physical Review B, 64*(22), 224205.
51. Cormier, L., Neuville, D. R., & Calas, G. (2000). Structure and properties of low-silica calcium aluminosilicate glasses. *Journal of Non-Crystalline Solids, 274*(1), 110–114.
52. Youssef, M., Pellenq, R. J. M., & Yildiz, B. (2011). Glassy nature of water in an ultraconfining disordered material: The case of calcium–silicate–hydrate. *Journal of the American Chemical Society, 133*(8), 2499–2510.
53. Korb, J. P., & Monteilhet, L. (2007). Microstructure and texture of hydrated cement-based materials: A proton field cycling relaxometry approach. *Cement and Concrete Research, 37*(3), 295–302.
54. Bordallo, H. N., Aldridge, L. P., & Desmedt, A. (2006). Water dynamics in hardened ordinary Portland cement paste or concrete: From quasielastic neutron scattering. *Journal of Physical Chemistry B, 110*(36), 17966–17976.
55. Kocks, U. F., & Mecking, H. (2003). Physics and phenomenology of strain hardening: The FCC case. *Progress in Materials Science, 48*(3), 171–273.
56. Liu, L., Jaramillo-Botero, A., Goddard, W. A., & Sun, H. (2012). Development of a ReaxFF reactive force field for ettringite and study of its mechanical failure modes from reactive dynamics simulations. *Journal of Physical Chemistry A, 116*(15), 3918–3925.
57. Bonnaud, P. A., Ji, Q., Coasne, B., Pellenq, R. J., & Van Vliet, K. J. (2012). Thermodynamics of water confined in porous calcium–silicate–hydrates. *Langmuir: The ACS Journal of Surfaces & Colloids, 28*(31), 11422.

58. Hou, D., Ma, H., Li, Z., & Jin, Z. (2014). Molecular simulation of "hydrolytic weakening": A case study on silica. *Acta Materialia, 80,* 264–277.
59. Zhu, T., Li, J., Lin, X., & Yip, S. (2005). Stress-dependent molecular pathways of silica–water reaction. *Journal of the Mechanics and Physics of Solids, 53*(7), 1597–1623.
60. Manzano, H., Durgun, E., López-Arbeloa, I., & Grossman, J. C. (2015). Insight on tricalcium silicate hydration and dissolution mechanism from molecular simulations. *ACS Applied Materials & Interfaces, 7*(27), 14726–14733.

Chapter 7
Molecular Dynamics Study on Cement–Graphene Nanocomposite

Nanotechnology has been utilized to improve the properties of the cement-based material. In previous chapters, the nanoscience of the cement hydrate has been investigated by means of molecular simulation. It provides valuable insights on the molecular structure, dynamics, and mechanical properties of cement hydrate at the nanoscale. These findings can help material design by nanoengineered method, and the nanomaterial such as the graphene-based material can be incorporated into the cement matrix and heal the defection in the cement hydrate, further improving the properties of the material. In this chapter, molecular dynamics was utilized to study the strengthening mechanism of GO sheets on the C–S–H gels.

7.1 Introduction

Graphene, the single-atom-thick sheet constructed by sp^2-bonded carbon atoms, have drawn much attention due to its numerous unique properties, including excellent carrier mobility, outstanding tensile strength, extremely large specific surface area, and high thermal conductivity [1]. During the past few years, graphene-based materials, such as graphene sheet and graphene oxide (GO), have been incorporated into the matrix of various materials for a wide range of functional applications: extraordinary high-strength polymer nanocomposites with graphene platelets [2], shape-memory and enzyme-sensing bio-materials incorporated by the graphene and amyloid fibrils [3] and high-performance thermally insulating materials developed by nanocellulose and GO [4]. In particular, the superior mechanical performance of graphene sheet (Young's modulus: 1 TPa, tensile strength 130 GPa) makes it an ideal candidate for advanced filler materials in cement-based composites [5]. Reinforcement of traditional cementitious materials with graphene can open up new pathways for improving the mechanical properties and durability of building materials, and can significantly reduce the consumption of traditional concrete materials. In respect to the sustainable development, it is of great importance and urgency to develop novel

© Science Press and Springer Nature Singapore Pte Ltd. 2020
D. Hou, *Molecular Simulation on Cement-Based Materials*,
https://doi.org/10.1007/978-981-13-8711-1_7

building materials, especially considering that the manufacture and application of concrete results in the release of 5–10% of global greenhouse gas each year [6].

However, difficulties in dispersing graphene and the high cost of production limit its widespread application in construction and building materials. As a graphene derivative, GO is a monolayer of sp^2-hybridized carbon atoms modified by a mixture of carboxyl, hydroxyl, and epoxy functionalities [7]. The oxygen functional groups, attached on the basal planes and edges of GO sheets, significantly alter the van der Waals interactions between the GO sheets and therefore improve its dispersion in water. The extraordinary mechanical properties, highly dispersible property in water, and lower cost, make GO a promising material for enhancing the mechanical properties of cement composites. Pan et al. [8] have shown that the addition of 1.0 wt% GO could simultaneously improve the strength and toughness of GO–chitosan composites. This improvement has been attributed to the enhanced nanofiller–matrix adhesion/interlocking arising from the wrinkled surface and the two-dimensional geometry of the graphene platelets. Lv et al. [9] showed that cement composites exhibited a remarkable increase in tensile strength (78.6%), flexural strength (60.7%), and compressive strength (38.9%) by incorporating 0.03 wt% GO. Pan et al. [10] found that the introduction of 0.05 wt% GO increased the compressive strength of GO–cement composite by 15–33% and the flexural strength by 41–59%, respectively. Lu et al. [11] demonstrated that 0.05 wt% GO led to 11.1 and 16.2% increase in compressive strength and flexural strength of cement paste, and also showed the effect of the hybrid GO/CNTs composites on the mechanical behavior of cement paste. Abrishami and Zahabi [12] studied reinforcing graphene oxide/cement composite with NH_2 functionalizing group and found that the flexural strength can be increased by 38.4% by compositing 0.1 wt% GO in the cement paste. They attributed the strength enhancement to the porosity decreasing and improvement of interfacial strength between C–S–H and GO. Furthermore, the empirical force field molecular dynamics simulation has been performed to study the molecular-scale energetic, structural, and dynamic properties of the interface between surface functionalized graphitic structures and calcium–silicate–hydrate (C–S–H) [13, 14]. The simulation work provided molecular insights on the H-bonds interaction between C–S–H and different functional groups in GO structures. However, the empirical force field fixed the bonds in the functional groups on the GO and cannot allow the chemical reactions between the C–S–H gel and GO. Hence, it is necessary to further investigate the interaction between GO and more realistic C–S–H model by reactive force field molecular dynamics.

Many studies have been conducted to combine graphene/GO with cement-based materials [15], but they mainly focused on improving the mechanical strength and transport behavior. The interaction mechanism between graphene, GO and C–S–H gel has not been understood comprehensively on the nanoscale level. In this chapter, reactive force field MD simulation will be utilized to investigate the C–S–H and graphene/GO interface properties, including surface energy, molecular structure, and cohesive strength, which provide scientific guidelines for enhancing macroscopic properties of cement-based materials. The comprehensive study of the nanointeraction, microstructure, and macro-behavior of graphene/GO-modified cement-based materials can help guide the sustainable design of materials using a fundamental approach.

7.2 Simulation Methods

7.2.1 Force Field

Molecular dynamics simulation, including the model construction and mechanical testing, was carried out by using the ReaxFF force field. The ReaxFF set of parameters employed in this study was created by merging the Si–O–H, C–O–H, and Ca–O–H sets developed by van Duin et al. [16, 17] and Manzano et al. [18].

7.2.2 Model Construction

The simulation of the crystal structure is based on the 11 Å tobermorite structure determined by Hamid in 1981 [19] that is taken as the C–S–H substrate. Cleavage of the tobermorite structure in the [001] direction leads to non-bridging oxygen (NBO, Si–O–). The silicate chains are assumed to be infinite with no defective part. The bridging oxygen (BO) and NBO from the silicate chains become the predominated elements in the channel. Some interlayer calcium atoms are also adsorbed near the silicate channels. In the simulation, the tobermorite super-cell contains $2 \times 3 \times 1$ crystallographic unit cells. The dimensions of the super-cell are $a = 22.32$ Å, $b = 22.02$ Å, $c = 22.77$ Å and $\alpha = 90°$, $\beta = 90°$, $\gamma = 90°$. After completion of cleaving the super-cell, three models with different interlayer graphene and GO are constructed.

The pristine graphite unit cell with lattice parameters $a = 2.46$ Å, $b = 4.26$ Å, $c = 3.4$ Å and $\alpha = 90°$, $\beta = 90°$, $\gamma = 90°$ was constructed, following previous research [14]. The cell was then replicated 9 times along the a-direction and 5 times along the b-direction, while the thickness remained as 3.4 Å. The graphene sheet, including 180 carbon atoms, was inserted into the interlayer region created by the cleaved C–S–H substrates. This initial model is defined as graphene_C–S–H (G_C–S–H). Additionally, the GO structure was created by attaching 18 functional groups (–OH) on the graphene sheet, with a surface coverage of 10%. According to different degrees of oxidation, the coverage ratio might vary and the 10% coverage of –OH groups is in the range of the typical surface treated carbon fiber materials published in Refs. [20, 21]. The GO confined by the C–S–H substrates construct the second model defined as GO_C–S–H. To consider the interaction between the GO and Al-rich C–S–H, the 16 Al atoms are introduced into the interlayer region. To ensure the charge balance, 24 interlayer Ca atoms are removed. The third model is defined as GO_C–A–S–H. At the very beginning, all the atoms in the graphene/GO were located 5 Å away from either substrate surface. The model construction procedure is schematically shown in Fig. 7.1.

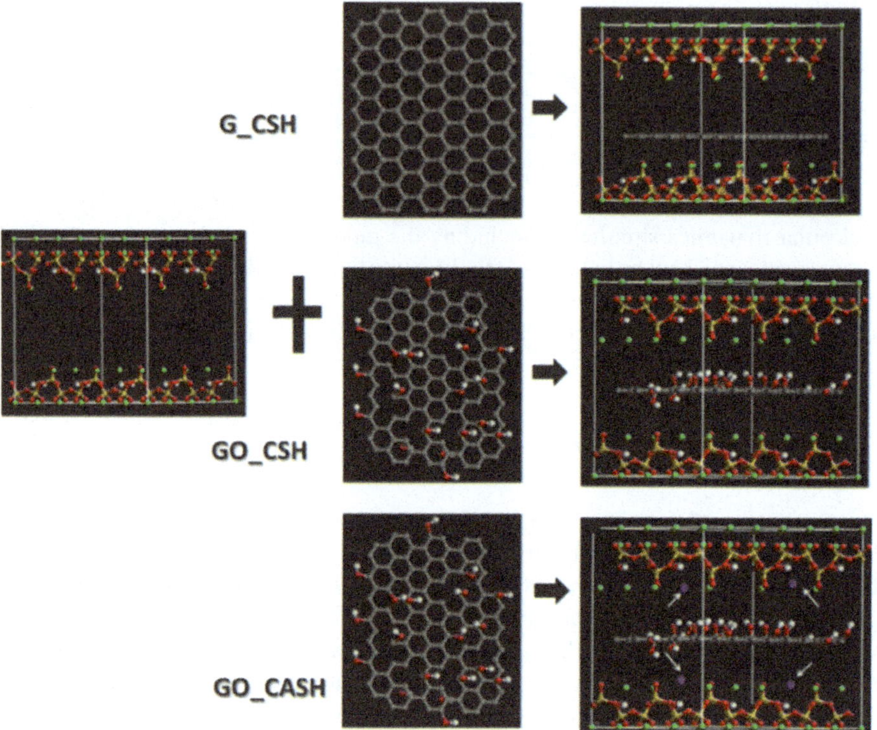

Fig. 7.1 Model construction of graphene/C–S–H, GO/C–S–H, and GO/C–A–S–H composite. The left figure represents cleaved tobermorite; three figures in the middle are G and GO sheets; right figures are the initial model of G/C–S–H, GO/C–S–H, and GO/C–A–S–H before molecular dynamics. The ball–stick styles represent the composite model. The red, green, gray, yellow, white, and purple balls represent oxygen, calcium, carbon, silicon, hydrogen, and aluminum atoms, respectively. The white-red stick represents the hydroxyl bond, the gray stick is the C–C bond, the yellow-red stick is the silicate bond

The trajectories of the atoms were calculated by the Verlet algorithm with a time step of 0.25 fs. Initially, the Nosé–Hoover thermostat algorithm was used to simulate the interaction between the C–S–H gel and graphene/GO in the NPT ensemble at 300 K and 1 atm for 250 ps. When the system reached equilibrium, it turned to the canonical ensemble at 300 K 100 ps. Finally, a further 1000 ps NPT production run was performed to obtain the atomistic trajectories for structural and dynamics analysis.

7.3 Molecular Structural Properties of Graphene/GO and C–S–H Model

7.3.1 Molecular Structure of Graphene/GO and C–S–H

Molecular structures of the composite of the C–S–H and graphene/GO are shown in Fig. 7.2. Correspondingly, Fig. 7.3 demonstrates the intensity profile of the Ca, Si, O, H, Al, and C atoms distribution along the z-direction. For the C–S–H structure, the silicate chains rooted in the calcium sheet form the calcium silicate layers. The directly connected silicate tetrahedron is defined as paired (P) and the silicate tetrahedron connecting two paired tetrahedrons is defined as the bridging tetrahedron (B). The water molecules, hydroxyl groups, and the interlayer calcium ions are located near the surface of the bridging tetrahedron or in the vicinity of the channel between neighboring bridging tetrahedrons. Differing from the initial structure of tobermorite 11 Å without hydroxyl groups, there are some silicate hydroxyl (Si–OH) and calcium hydroxyl (Ca–OH) groups produced after the reactive simulation. Some of the water molecules near the layer are dissociated into H^+ and OH^-. While the H^+ is associated with the non-bridging oxygen atoms in the bridging tetrahedron, the OH^- is bonded with the interlayer calcium atoms [22]. Both the non-bridging sites and adsorbed calcium atoms contribute to the hydrophilic characteristic of the C–S–H surface [23]. Interestingly, it can be clearly observed from Fig. 7.2a, b that the GO/C–S–H interface has more Si–OH groups than that of the graphene/C–S–H one. It indicates that proton transferring might occur between the C–S–H and GO, and is further explained in the following section. Despite of the presence of the G/GO in the interlayer region, the silicate chain with infinite length in the tobermorite crystal maintained the calcium silicate structure in the ordered layer state. In the real C–S–H gel, the calcium silicate layer might be disturbed to some extent as the silicate chains turn defective.

The graphene and GO confined in the interlayer region of the C–S–H gel demonstrates different morphologies. In the graphene, the carbon atoms, orderly arranged in the xy plane, are distributed away from the C–S–H surface. No extension along the z-direction can be observed in the graphene/C–S–H interface in Fig. 7.2a. As shown in Fig. 7.3b, the sharp peak of the C atoms with a few overlapping with hydrogen atoms in the C–S–H gel, also confirms the hydrophobic nature of the graphene sheet. On the other hand, as shown in Fig. 7.2b, the protruding hydroxyls on both sides of the GO point toward the C–S–H interface. The C–C bonds in the GO are stretched in the z-direction and the graphene sheet is disturbed to some extent. In Fig. 7.3b, the distribution of hydrogen atoms growing through the interlayer region suggests improved interaction between the C–S–H and GO. The oxygen-containing groups can weaken the repulsive effect between the single-atomic carbon layer and the C–S–H gel and reduce the interlayer distance, as shown in Fig. 7.3d. Furthermore, the aluminum atoms in the interface region can greatly change the GO structure greatly. As shown in Fig. 7.2c, the ordered two-dimensional GO sheets are transformed to the amorphous three-dimensional carbon network structures, losing the

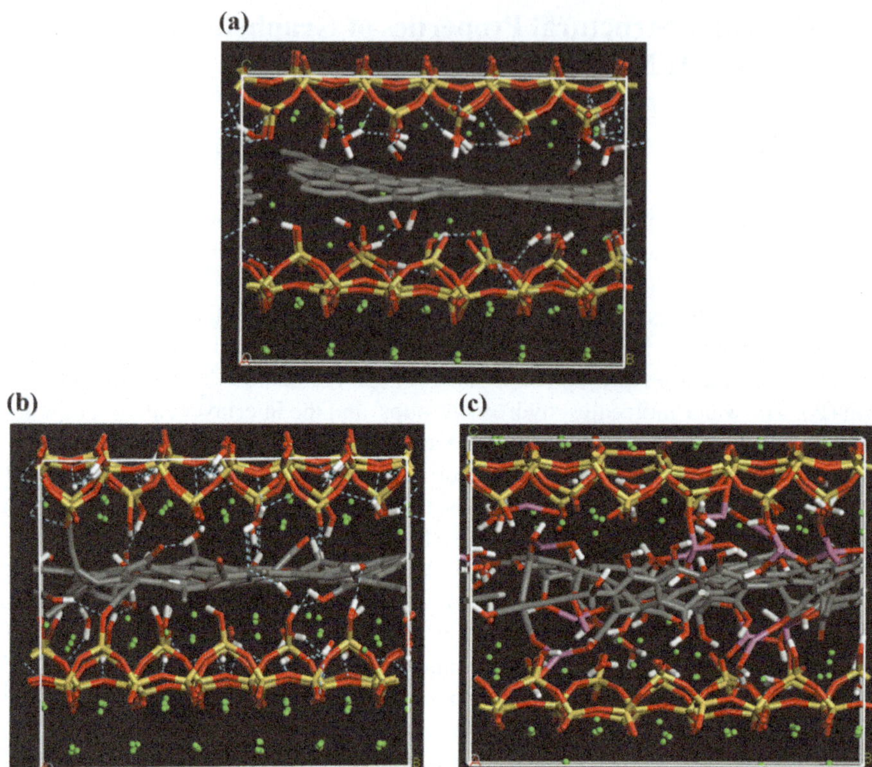

Fig. 7.2 Molecular structure of the **a** G/C–S–H; **b** GO/C–S–H; **c** GO/C–A–S–H in the equilibrium state after molecular dynamics simulation. The meanings of the ball and sticks with different color are illustrated in Fig. 7.1 caption

repetitive pattern of hexagons for the carbon atoms. The broad distribution of the carbon atoms in Fig. 7.3c implies an expansion of the interlayer region. It can be observed that the aluminum atoms, forming the Si–O–Al–O–C connection, play the role in bridging the silicate tetrahedron and the carbon hexagons. This enhances the interaction between the C–S–H and GO. A healing effect for the Al atoms in the defective silicate chains has also been found in the C–A–S–H gel [24]. Because the GO sheet is strongly stretched by the neighboring atoms in the z-direction, the interlayer distance increases to around 18.8 Å, as seen in Fig. 7.3d.

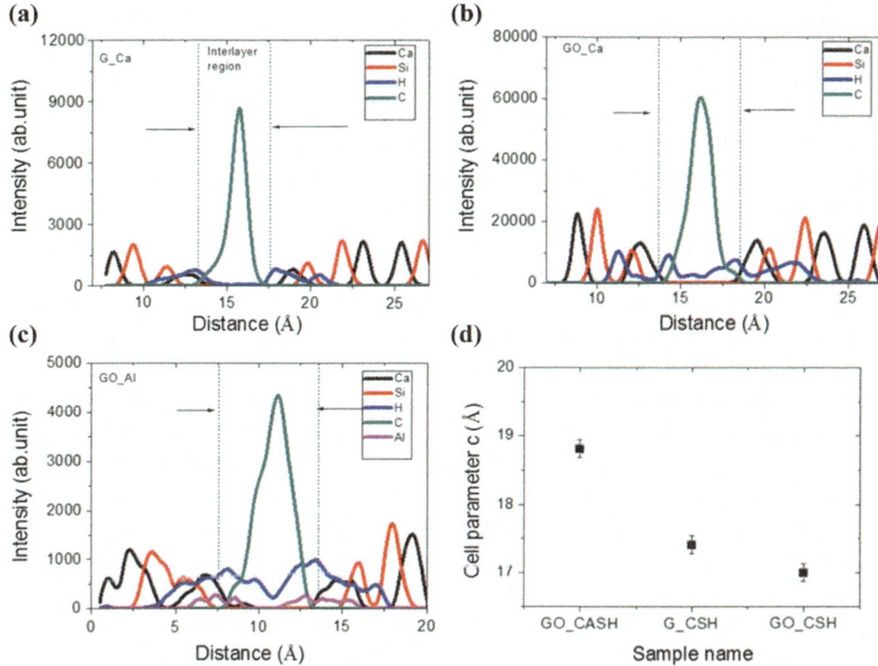

Fig. 7.3 Density profile for the atoms along z-direction **a** G/C–S–H; **b** Go/C–S–H; **c** Go/C–A–S–H; **d** the crystal parameter c (cell size along z-direction)

7.3.2 Local Structure of the Graphene and GO in the Interlayer

The spatial correlation for the carbon atoms in the graphene/GO can be further characterized by the radial distribution function of C–C and C–O. As shown in Fig. 7.4a, in the short range, there are three peaks with decreasing intensity located at 1.44, 2.52, and 2.91 Å in the RDF of the C–C for graphene, representing the C–C bond length and typical diagonal distances in the same regular carbon hexagon. The second and third peaks emerge to a broader one in the C–C spatial distribution for the GO, mainly resulting from distortion of the carbon hexagons and defects in the structure. In the medium range, in the graphene, the local maxima of RDF are positioned at 3.81, 4.33, and 5.07 Å, indicating the strong spatial correlation between neighboring carbon hexagons. By geometric calculation, the relative atomistic positions in two different hexagons also confirm that the carbon atoms are in the same plane in a range around 5 Å. However, the weaker peak intensities of the GO suggest that the confined GO cannot remain in a flat state even in the medium range. The neighboring hexagons rotate around their shared edge and the torsion results in the wrinkles of the GO structures, which can be clearly observed in Fig. 7.2b, c. Furthermore, the intensity peaks of the graphene can extend to around 9 Å, implying that the structural order can persist for a long distance. This can be attributed to the two-dimensional extension of the graphene structure. In the long range (>6 Å), no obvious spatial correlation can be observed in the GO that shows amorphous branch structures being ultra-confined in the interlayer region.

The RDF of the C–O atoms can further describe the local structures of the oxygen-containing groups grafted in the graphene. As expected in Fig. 7.4b, the RDF intensity remains at a zero value until the distance reaches 2.5 Å in the graphene and C–S–H interface. In the range from 2.5 to 10 Å, no intensity maximum is detected in the RDF, implying no spatial correlation with the neighboring oxygen atoms in the silicate hydroxyl and water molecules near the surface of C–S–H. On the other hand, the oxygen atoms bonding to the GO result in pronounced peaks positioned at 1.41, 2.54, and 3.76 Å in the RDF, which reflect the direct C–O covalent bond, indirect C–C–O and the C–C–C–O connections in the single benzene-ring like structure. It should be noted that a C–O RDF from 1.3 to 1.8 Å demonstrates a bimodal distribution in Fig. 7.4c, indicating two different C–O bond types. As shown in Fig. 7.4d, the first type is C–O with a shorter length around 1.37 Å and the second type is the hydroxyl groups connected with carbon atoms, with bond length around 1.43 Å. Additionally, while the first peak height in RDF for the GO/C–S–H is quite lower than that of the second one, the heights of the double peaks for the GO/C–A–S–H show little difference. It means that the C–OH groups occupy a predominant percentage in the GO/C–S–H, yet in the GO/C–A–S–H, the ratios of C–O and C–OH are similar. Since no C–O groups were prepared at the beginning of the model construction, they were all produced after the reactive simulation process. The chemical reaction pathway can be illustrated in the following sequence in Fig. 7.4e: the interlayer calcium or aluminum ions diffuse and approach the C–OH groups due to electronic attraction; the electronegativity of the oxygen atoms are shared with neighboring Ca and Al atoms and the O–H bond is stretched; the C–OH dissociates into H atoms and the C–O group; the de-bonded hydrogen atom diffuses and associates with the non-bridging oxygen atom in the silicate tetrahedron and forms the Si–OH connection. Besides, the free hydrogen atoms can also associate with the calcium hydroxyl groups, producing water molecules. This reaction explains the observation that more silicate tetrahedrons are protonated in the C–S–H surface in the presence of GO. In the current simulation, –OH groups on the GO surface can dissociate into H^+ and $–O^-$, resembling the ionization in phenol where there is an equilibrium between the –OH, H^+, and $–O^-$ groups in an aqueous solution [25]. Under the conjugated structure of un-oxidized benzene rings, the $–O^-$ groups could exist in a stable manner by sharing the electrons. The ionization degree can be improved by the reaction between H^+ and hydroxyl groups in the C–S–H gel. Compared with calcium atoms, the aluminum species are more likely to form stable Al–O–C covalent bonds, which can accelerate the proton transfer between the GO and C–S–H surface. In previous research, Sanchez et al. [14] investigated the interaction between functionalized graphitic structures and calcium silicate hydrate and found that the affinity to C–S–H was closely related with the polarity of the functional group. In current study, the de-protonation reaction can contribute to the high polarity of the GO system, which to some extent enhances the interaction with the C–S–H structures.

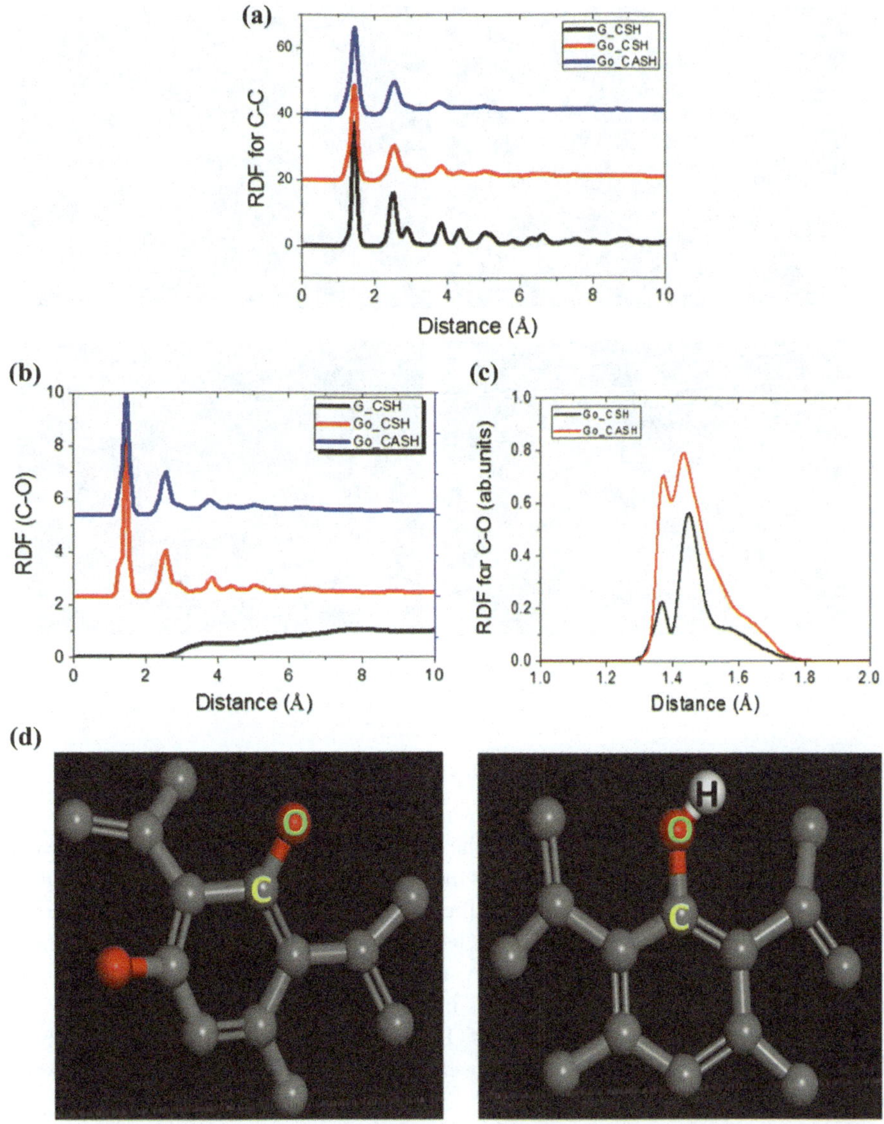

Fig. 7.4 **a** Radial distribution function (RDF) for the C–C atoms in graphene/GO systems; **b** RDF for the C–O atoms in graphene/GO systems; **c** short-range RDF for C–O atoms in GO systems; **d** molecular structure of the C–O and C–OH bonds in GO; **e** schematically description for the C–OH dissociation. The meaning of different colored ball and sticks is in Fig. 7.1 caption

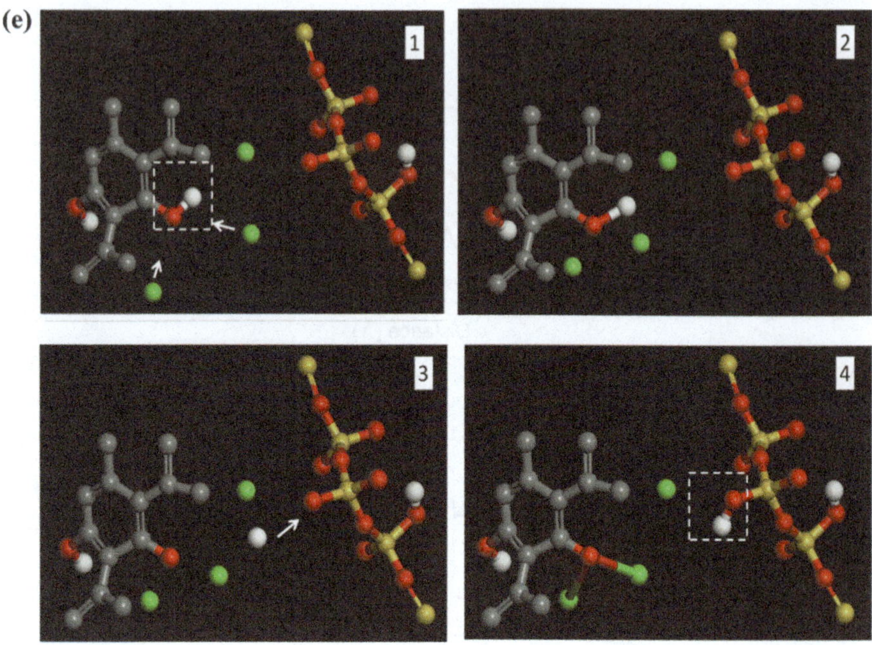

Fig. 7.4 (continued)

7.3.3　Local Structure of Interlayer Ca and Al Ions

The molecular structure of the GO in the vicinity of the C–S–H surface is greatly affected by the ions and hydroxyl groups in the interlayer region. Figure 7.5a exhibits the RDF for the Ca_w (calcium atoms in the interlayer region) and O_c (oxygen atoms bonded to carbon) in the samples of GO/C–S–H and GO/C–A–S–H. The sharp peaks in both cases are located at 2.67 Å, implying a stable Ca–O connection in the interface. Meanwhile, the oxygen atoms in the samples are further categorized into oxygen atoms in the C–S–H gel (O_s) and O_c. In Fig. 7.5b, strong correlations for both the $Al–O_c$ and $Al–O_s$ have also been observed at distance 1.92 Å, which represents the Al–O covalent bond.

The local structures of the Al and Ca atoms are also analyzed by the coordination numbers (CN). As shown in Fig. 7.5c, on average, the Ca atoms have around 5.9 nearest oxygen neighbors, including 4.4 O_s and 1.5 O_c. A CN of nearly 6 has been widely found in silicate glass [26] and cement-based materials [27] in previous investigations. The 6-coordinated Ca–O clusters are more likely to form disordered octahedrons. On the other hand, 2.2 O_s atoms and 1.8 O_c atoms contribute to 4 coordinated atoms of Al species. Al and O atoms aggregate to build the aluminate tetrahedron. The aluminate tetrahedron structures occupy predominant percentage in the C–A–S–H gel, which matches well with recent NMR characterization on the silicate–aluminate structures [28]. Hou et al. [29] and Qomi et al. [24] studied the local

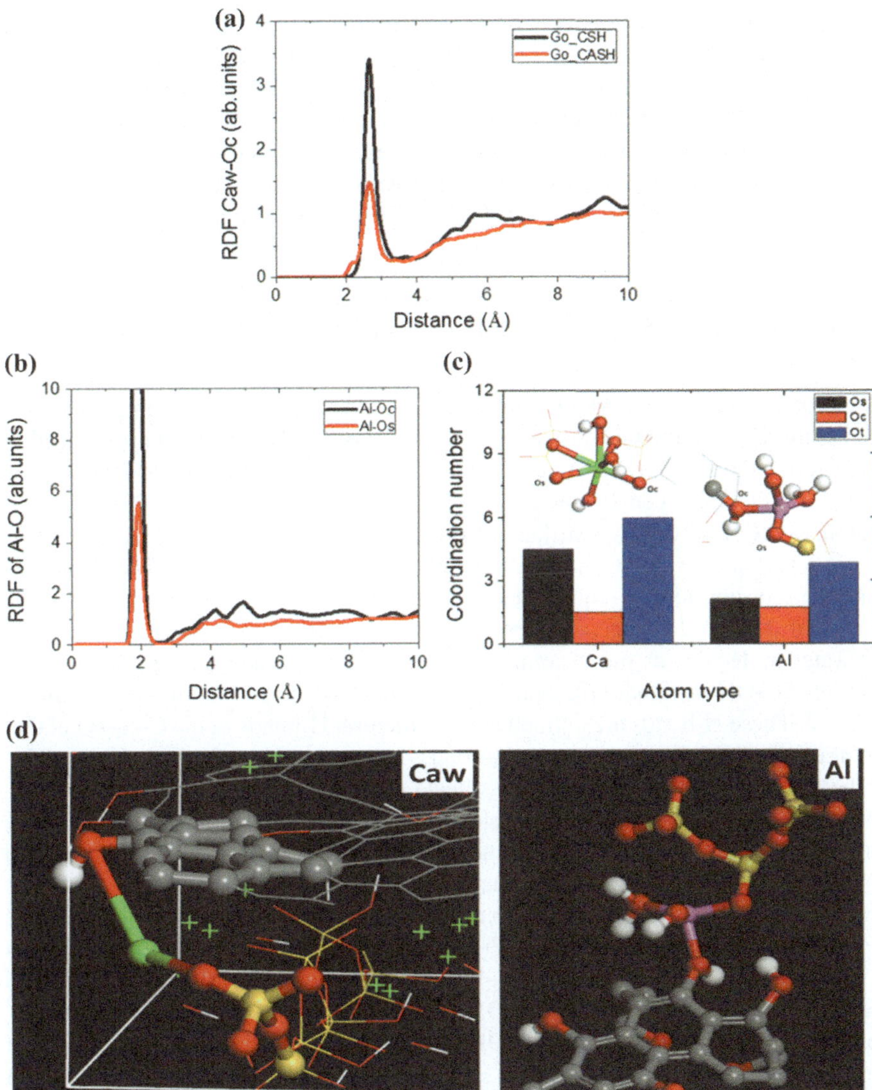

Fig. 7.5 **a** Radial distribution function for Ca–O; **b** RDF of Al–O; **c** coordination number of Ca–O and Al–O, and local molecular structure of Al–O and Ca–O cluster; **d** molecular structure of O_s–Al–O_c and O_s–Ca–O_c connecting the neighboring graphene oxide and C–S–H gel. The meaning of different colored ball and sticks is in Fig. 7.1 caption

structure of Al species presents in the interlayer region of C–S–H gel by molecular dynamics. Both studies found that Al species can associate with non-bridging sites of the defective silicate chains, constructing five- and sixfold polyhedrons. Since the infinite long chains in the tobermorite 11 Å have fewer non-bridging atoms, the CN of Al species reduces to around 4 in the current research. As illustrated in Fig. 7.5d, the counterions in the vicinity of the C–S–H surface play an essential mediating role in bridging the coordinated atoms in the C–S–H structure and GO.

7.3.4 The Local Structure of Water and Hydroxyl Groups

In addition to Ca–O and Al–O bonds, the H-bonds are important components for linking the C–S–H and GO structures. The RDFs of O_c–H and O_s–H are utilized to characterize the structural feature of the H-bonds network in the interfacial region. In the interface between graphene and C–S–H, as exhibited in Fig. 7.6a, the first peak in the RDF of O_s–H, distributed from 1.4 to 2.0 Å, reflects the H-bonds inter-action between the silicate chains and neighboring water molecules. The H-bond is defined according to two requirements: bond length is smaller than 2.45 Å and the minimum donator–hydrogen–acceptor angle is 150° [30]. As shown in Fig. 7.6b, the water molecules or the decomposed hydroxyls can either accept the H-bonds from the Si–OH or donate the bonds to the bridging oxygen atoms in the silicate skeleton. These H-bonds are defined as the structural H-bonds in the C–S–H gel that are dramatically different from the H-bonds between neighboring water molecules in the bulk water solution [18]. The silicate chains provide a large number of H-bonds accepting sites for the water molecules, which can further restrict the mobility for hydroxyl groups. On the other hand, as compared with the shallow distribution for the H-bonds in the silicate chains, the first peak for the RDF of O_c–H shifts towards a longer distance with a broader distribution from 1.48 to 2.62 Å. Three types of H-bonds can be observed in Fig. 7.6c: the de-protonated C–O accepts bonds from the hydrogen in Si–OH (C–O—HO–Si); the neighboring hydroxyl rooted in the GO interact with each other (C–OH—C–OH); the carbon hydroxyl donates H-bonds to calcium hydroxyl (C–OH—OH–Ca). Three types of H-bonds show different bond lengths: C–O—HO–Si (1.77 Å) < C–OH—OH–Ca (1.99 Å) < C–OH—C–OH (2.36 Å). It indicates that the H-bonds properties are greatly influenced by the local environment. A similar H-bonds structure can also be observed in the Al-rich region. As exhibited in Fig. 7.6d, the hydroxyl in the silicate tetrahedron accepts H-bonds to neighboring aluminate tetrahedron, and meanwhile acts in the role of H-bonds donor attracting the C–OH groups. In this respect, the first and third types are important chemical bonds for bridging the C–S–H gel and GO.

Dipole moment distribution of a single water molecule has been employed to study the interaction between water molecules and their C–S–H or graphene/GO substrates quantitatively. The dipole moment is sensitive to the local environment of the water molecules: while the hydrophilic confinement contributes to a higher value of dipole moment, the hydrophobic confinement tends to a lower value [23].

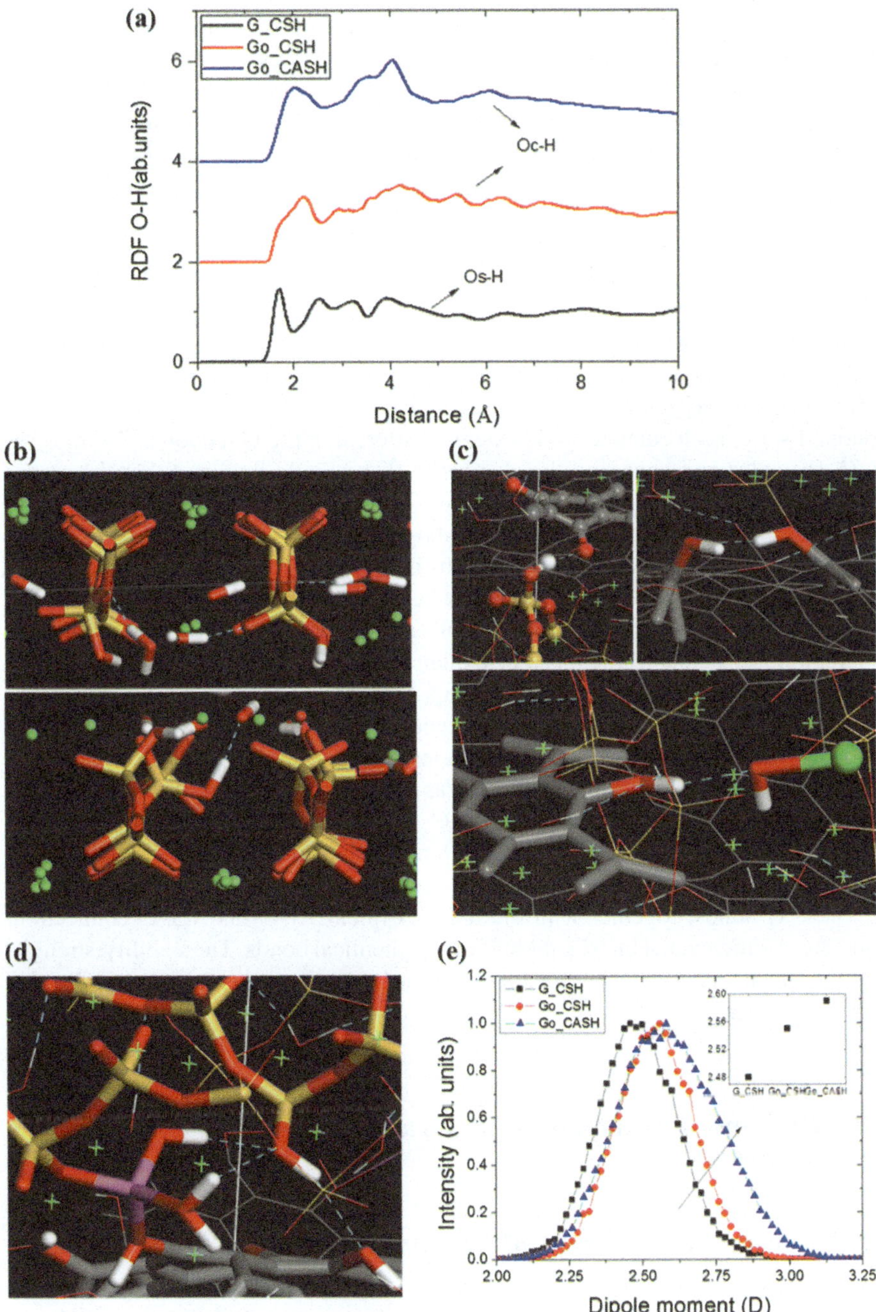

Fig. 7.6 **a** RDF of O_s–H in the G/C–S–H and O_c–H in Go/C–S–H, Go/C–A–S–H; **b** molecular structure of the structural H-bonds network; **c** molecular structure of three types of H-bonds between C–S–H and graphene oxide; **d** H-bond network in the Al-rich region; **e** dipole moment distribution for water molecules in graphene/C–S–H, GO/C–S–H and GO/C–A–S–H. The meaning of different colored ball and sticks is in Fig. 7.1 caption

The dipole moment has been calculated according to the methodology mentioned by Manzano et al. [18]. In previous ReaxFF simulation on the bulk water solution [31], the dipole moment for bulk water was found to be around 2.15 D and can be taken as reference. It can be observed from Fig. 7.6e that the average values for the dipole moment in all cases are significantly larger than 2.15 D. It indicates that the environment in which the water molecules located is hydrophilic in both graphene/C–S–H and GO/C–S–H.

As compared with graphene/C–S–H, the dipole moment distribution of water in GO/C–S–H shifts towards a larger value, with the mean value rising from 2.48 to 2.54 D. It confirms the hydrophilic nature of the functionalized C–OH groups in GO. When the water molecules are confined between the GO and C–S–H, the upshift for the dipole moment distribution is mainly attributed to the electronic field induced by the oxygen-containing groups and silicate framework that polarize the confined water molecules to some extent by stretching the O–H bonds, bending the H–O–H angles and changing the charges [32]. It should be noted that the carbon basal sheet still consists of un-oxidized benzene rings, which are hydrophobic. The hydrophilic oxygen-containing groups and hydrophobic benzene rings contribute to the amphiphilic property of GO and the am-hydrophilicity is influenced by the local PH and chemical compositions [33]. It can be observed that there is a further positive shift toward large values of dipole moment as the Al atoms are present in the interfacial region. This enhancement can be explained by the following two reasons. First, the effect of the interlayer cations on the dipole moment has been investigated in previous research, and it was found that Ca^{2+} and Al^{3+} ions contribute to the hydrophilic nature of the C–S–H gel [23] and Al species provide more positive charges for the polarizing water to balance them. Second, as discussed in the previous section, the Al species result in proton dissociation from C–OH, increasing the polarity of the functional group in GO. A GO framework with more polarity also leads to a hydrophilic feature.

These H-bonds, together with the ionic–covalent Al–O and Ca–O connections construct a dimensional interfacial network of chemical bonds. The stability, strength, and the mechanical contribution for different bonds is discussed in the following sections.

7.4 Dynamic Properties of the Graphene/GO and C–S–H Model

7.4.1 Dynamic Properties of Carbon Atoms

Figure 7.7 records the MSD of carbon atoms in the graphene/GO structure during 100 ps. As shown in Fig. 7.7a, in the graphene/C–S–H system, MSD continues increasing to 9 Å^2 during the initial 3 ps; subsequently drops to 2.6 Å^2 and fluctuates at this value until 30 ps; finally jumps to around 12 Å^2 and reaches a plateau in the last

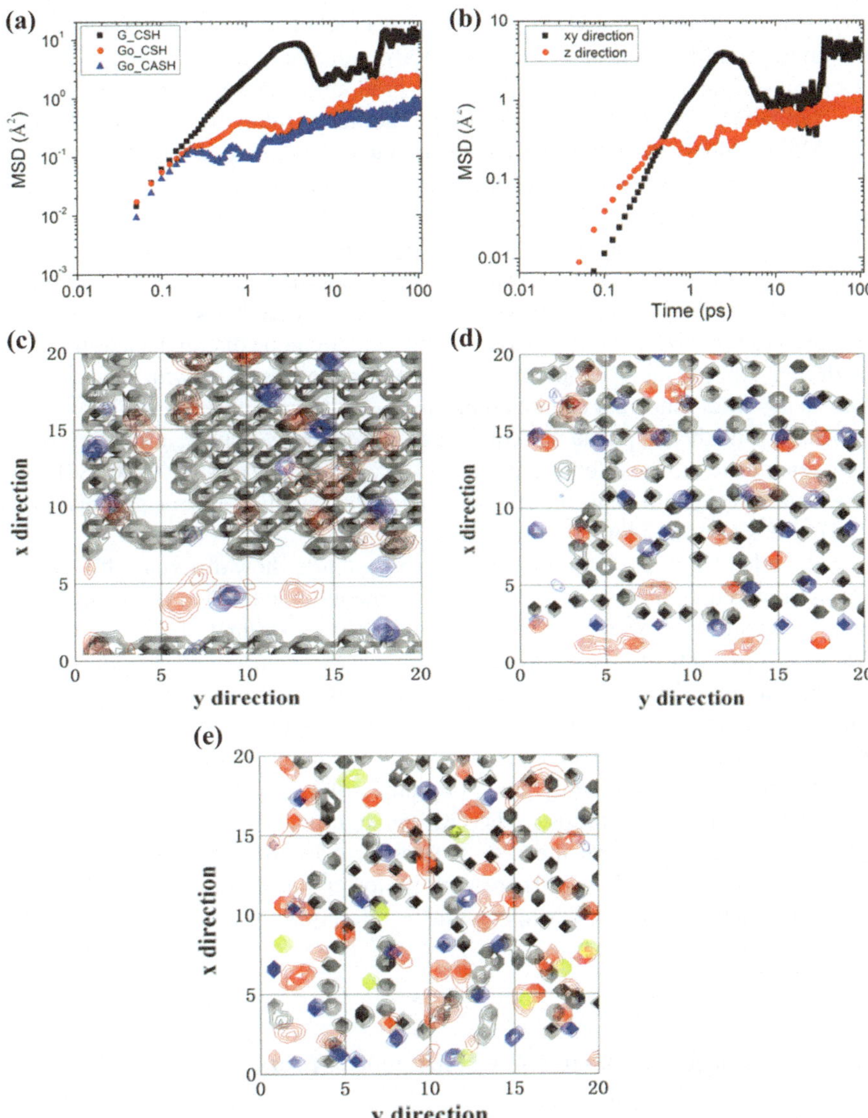

Fig. 7.7 a Mean square displacement for *C* atoms in graphene/C–S–H, GO/C–S–H and GO/C–A–S–H; **b** MSD$_{xy}$ and MSD$_z$ for C atoms in graphene/C–S–H; **c** trajectories of the atoms in *xy* plane in graphene/C–S–H; **d** in GO/C–S–H; **e** in GO/C–A–S–H

60 ps. Figure 7.7b shows the MSD_{xy} and MSD_z for carbon atoms in graphene/C–S–H. The variation of MSD_{xy} is almost simultaneous with that of MSD, implying that the movement of C atoms in the xy plane plays a predominant role. On the contrary, the ultra-confinement restricts the mobility of C atoms in the z-direction, resulting in lower MSD_z values (<1 Å^2). The dramatically large fluctuation of the MSD between 2.6 and 12 Å^2 reflects that the graphene oscillates in the domain of 3 Å, which is the typical size of a single-carbon-hexagon cell.

The contour map of the atomistic positions in the xy plane is shown in Fig. 7.7c in order to better understand the dynamic behavior of the C atoms in the nanometer channel. The trajectories of neighboring C atoms completely overlap and only the hexagon-like organization with a cavity in the center can be distinguished. It means that during 100 ps, the C atoms frequently exchange positions with their bonded atoms in the short range but cannot escape the cage in the middle range. The swinging behavior of the graphene structure in the xy plane is mainly attributed to the heterogeneous repulsive effect from the bottom and top surfaces for the C–S–H gel. Since no bonds are built in the interfacial region, the unstable graphene cannot remain in a fixed position for a long time.

On the other hand, the C atoms in the graphene oxide show different MSD evolution as compared with those in graphene. After the jump in the first 0.3 ps, the MSD slowly climbs to 1–2 Å^2 with slight fluctuations. The extremely low MSD values, even at 100 ps, indicate that C atoms cannot escape from the cage built by their coordinate atoms. The trajectories for the C atoms in GO are shown in Fig. 7.7d, e. Almost all the carbon atoms are distributed separately, without any overlapping with other neighboring atoms. Hence, they can only vibrate or rotate at fixed positions. The discrepancy between the dynamic properties of C atoms in graphene and GO indicates that functional groups in the GO can significantly improve the chemical stability of the C atoms near the C–S–H surface. Additionally, during 100 ps, the MSD values of C in GO/C–A–S–H are slightly slower than those of C in GO/C–S–H. It suggests that the presence of Al species in the interlayer region can further stabilize the bonding with GO.

7.4.2 Dynamic Properties for Hydrogen Atoms

The MSDs of different atoms in GO/C–S–H have been calculated and are shown in Fig. 7.8a. According to the MSD values at 100 ps, the mobility of atoms is ranked in the following order: H > C > Ca > Si. MSD values below 1 Å^2 for Si and Ca atoms indicate that the calcium silicate sheet plays skeletal role in stabilizing the whole structure. In the graphene oxide, the C atoms can be categorized into "immobile" atoms that are directly bonded with C–S–H and "mobile" ones, away from the interfacial chemical bonds. Compared with the stable C–S–H skeleton, the slightly higher MSD values of the C atoms are mainly attributed to the movement of the "mobile" atoms. In the interlayer region, the highest MSD values of the H atoms suggest that H atoms are the most unstable species in the C–S–H gel. Not only can some interlayer

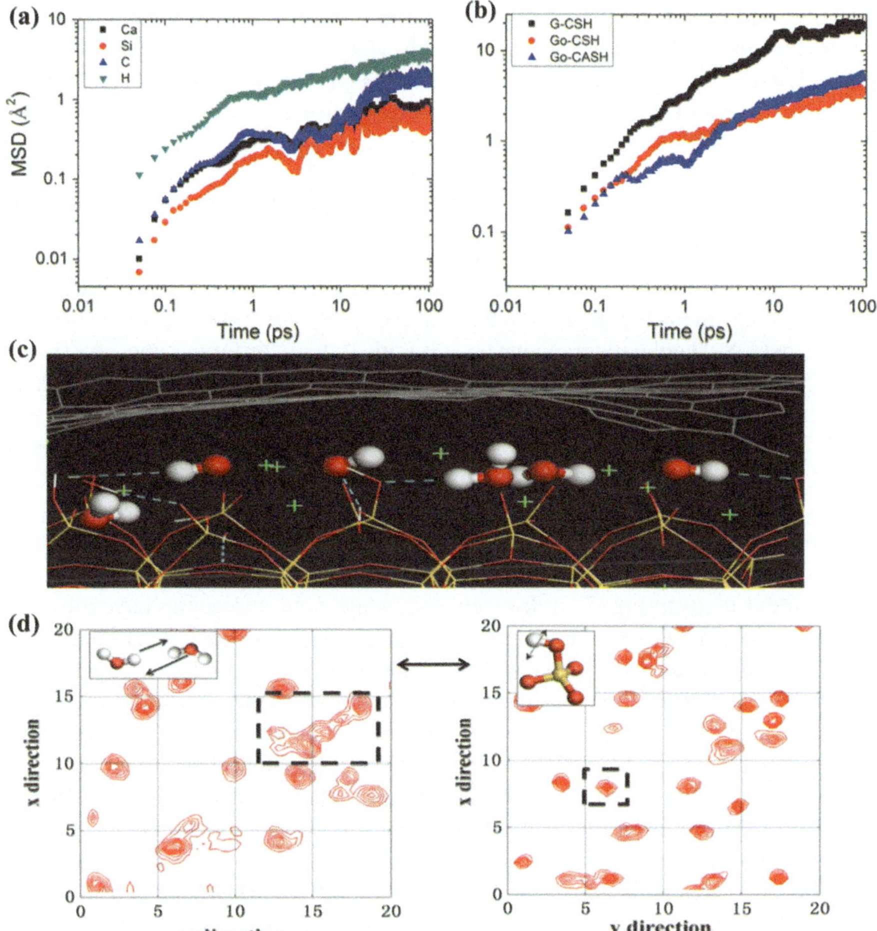

Fig. 7.8 **a** MSD for the Ca, Si, C, and H atoms in GO/C–S–H; **b** MSD for the H atoms in graphene/C–S–H, GO/C–S–H, and GO/C–A–S–H; **c** schematically description of water transport between graphene and C–S–H; **d** the trajectories of hydrogen atoms in xy plane for graphene/C–S–H (left) and GO–C–S–H (right). The meaning of different colored ball and sticks is in Fig. 7.1 caption

water molecules or hydroxyl groups diffuse in silicate channel, but the protons can transfer from C–OH to the Si–OH, as discussed in the previous section.

Additionally, as shown in Fig. 7.8b, the MSD values of the hydrogen atoms show large discrepancy for the water molecules or hydroxyl groups confined by graphene and GO. In the graphene, during 100 ps, the MSD continuously increases to around 20 Å^2, that is more than five times the value in GO. As seen in Fig. 7.8c, the dynamic behavior of the water molecules can be considered as the single-file water transport between hydrophobic graphene and the hydrophilic C–S–H substrate. Meanwhile, the trajectories of the hydrogen atoms, shown in Fig. 7.8d, indicate that some water molecules can diffuse in the xy plane of the graphene as far as 8 Å, but most of hydroxyl groups near the GO surface can only vibrate at the original positions.

The dynamic properties of the hydrogen atoms can be further investigated by the self-part of the van Hove function that is expressed in Eq. (7.1)

$$G_s(r, t) = \frac{1}{N} \sum_{i=1}^{N} \langle \delta(r - |r_i(t) - r_i(0)|) \rangle \tag{7.1}$$

The physical meaning of $2\pi r G_s(r, t)$ is the probability that particle i has moved a distance r in time t. Figure 7.9a, b demonstrates the van Hove functions of water confined in C–S–H gel with graphene and GO substrates at different times. As shown in Fig. 7.9a, at short timescale ($t < 1$ ps), the sharp peak located at 0.5 Å means there is a restriction effect from the cage. The average displacement shifts toward 0.9 Å at time 2.5 ps. At 25 ps, a clear intensity increase can be observed at 2 and 4 Å in $G_s(r, t)$. While the former length is a typical value for OH bonds vibrating at fixed Si–O$^-$ sites or the rotation around a fixed axis, the latter one is slightly larger than the length of one water molecule (~3 Å). Another intensity peak at 45 ps indicates that the hydrogen atoms can travel as far as 8 Å, which is consistent with the observation in the contour maps in Fig. 7.6d. It means that some water molecules have diffusive behavior between two neighboring silicate tetrahedrons at very short simulation times. The long distance diffusion is not possible for the water molecules confined in C–S–H gel is so short a time, due to strong attraction from the non-bridging oxygen sites and the over-solvated counterions [22], and the obstacles from the protruding bridging silicate tetrahedron [34]. Nevertheless, the graphene sheet does contribute to the transport of hydrogen atoms. Previous experimental study [35] and molecular simulation [36] proposed that water transport in the carbon nanotubes (CNTs) or in the vicinity of graphene is a fast phenomenon. The velocity accelerating effect of CNTs and graphene is due to the less friction interior surface that contributes to the super-lubricity for the water transport [37]. In this respect, the transport process for the water molecules is described as the following circle: molecules escape from the energy barrier of the C–S–H substrate, diffuse rapidly near the graphene surface and return to the C–S–H substrate. On the other hand, as time evolves, the shoulder located around 2 Å gradually increases in the $G_s(r, t)$ distribution for the movement of H atoms near the GO. The H atoms accumulation in the short distance domain implies that the aluminate–silicate chains, calcium ions, and C–OH groups construct the cages, strongly preventing the freely diffusion of the H atoms.

7.4.3　Time Correlation Function for Chemical Bonds

In the molecular structural analysis mentioned above, the H-bonds, Ca–O$_c$, and Al–O$_c$ are important connections that link the GO and C–S–H gel, and it is necessary to explore the strength and stability of the chemical bonds. The time corre-

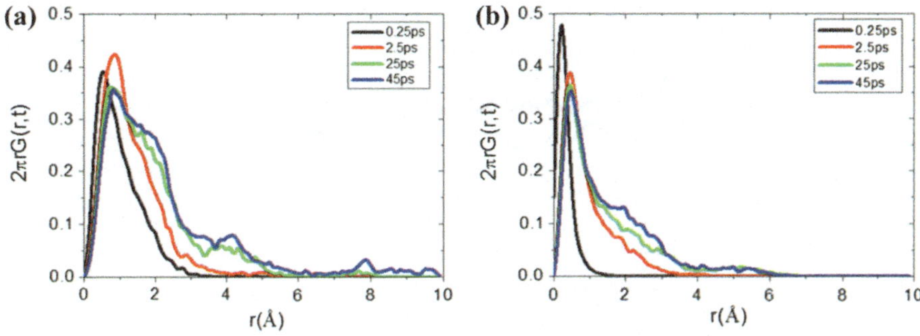

Fig. 7.9 $G_s(r, t)$ distribution of H atoms in **a** graphene/C–S–H; **b** GO/C–S–H at different simulation time from 0.25 to 45 ps

lated function (TCF) is utilized to describe dynamical properties of various chemical bonds.

As shown in Fig. 7.10a, with progressively increasing time, the TCF of H-bond in G and Go gradually reduces to 0.6 and 0.7, respectively. As compared with the G, the H-bond stability is enhanced to some extent in the Go system. This matches well with small MSD values of the hydrogen atoms near the Go surface that remain in fixed positions for a long time. Furthermore, the H-bonds can be divided into those connected with oxygen atoms in the C–S–H gel (O_s–H) and those with oxygen groups in the GO sheet (O_c–H), as mentioned in the above section. As shown in Fig. 7.10b, c, the bond strength of O_c–H is stronger than that of O_s–H in both GO–C–S–H and GO–C–A–S–H, in respect of bond stability. It is worth noting that the H-bond lifetime of the C–OH or C–O$^-$ groups is longer than that of the H-bonds of the structural water molecules. In addition, it was also found that the H-bond lifetime in the GO system was favored by the increasing of the GO polarity in previous molecular simulation [14]. On the contrary, according to the reduced value of TCF in Fig. 7.10b, the bond strength of Ca–O_c is weaker than that of Ca–O_s. Combined with the coordination analysis in the structural section, O_c only occupies one-sixth of the nearest neighbors to the Ca atoms. The bond stability of Ca–O_c is more likely influenced by other coordinates. Differing from the continuously reducing trend in TCF for the Ca–O and H-bonds, the TCF for the Al–O bonds remains constant during the simulation time, suggesting the strong strength of the covalent bonds. It resembles the aluminate–silicate skeleton in the C–A–S–H gel or in some zeolite crystals [24]. In summary, based on bond strength, the chemical bonds connecting the C–S–H and Go structures rank in the following order: Al–O > Ca–O_s > Ca–O_c > O_c–H > O_s–H. These chemical bonds play an essential role in withstanding the loading. The mechanical contribution of the different bonds is discussed in the following section concerning mechanical testing.

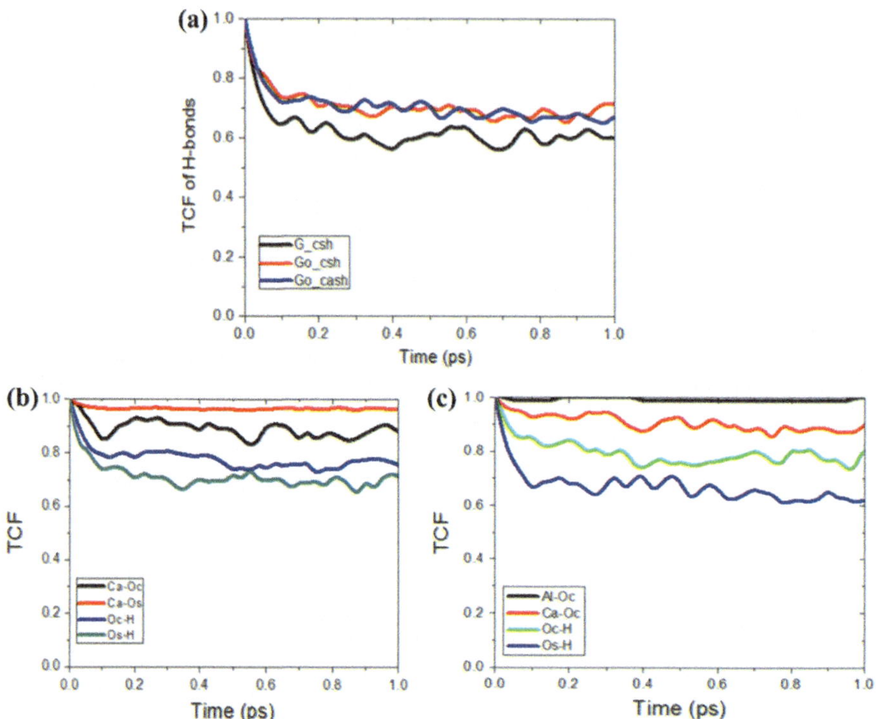

Fig. 7.10 a Time correlated function for H-bonds in graphene/C–S–H, GO/C–S–H, and GO/C–A–S–H. **b** TCF for different bonds in GO/C–S–H; **c** TCF for different bonds in GO/C–A–S–H

7.5 Reinforcement Mechanism of G/Go on C–S–H

The stress–strain relation can distinguish the different mechanical behaviors of the graphene/GO reinforced C–S–H gel under tension loading. As shown in Fig. 7.11, the stress fluctuates around 0 GPa with the increasing tensile strain for the graphene/C–S–H sample. As expected, the graphene without chemical bonding with the surrounding C–S–H substrates provides a little mechanical contribution. This is also due to the instability of the graphene confined in the interlayer region. Without bonding with C–S–H, the reinforced role of graphene on cementitious materials is quite limited in respect of some physical interactions such as the enhancing friction and filling the large voids.

On the other hand, as shown in Fig. 7.11 for the GO/C–S–H sample, the stress continues increasing to 1.5 GPa as the strain reaches 0.1 Å/Å, and subsequently reduces slowly in the post-failure regime. The tensile strength is within the range of previous findings on the cohesive force in C–S–H gel at the nanoscale (~3 GPa) [38]. The functional groups C–OH and –C–O bridge the C–S–H substrate and GO together to resist the mechanical loading rather than the two independent species in

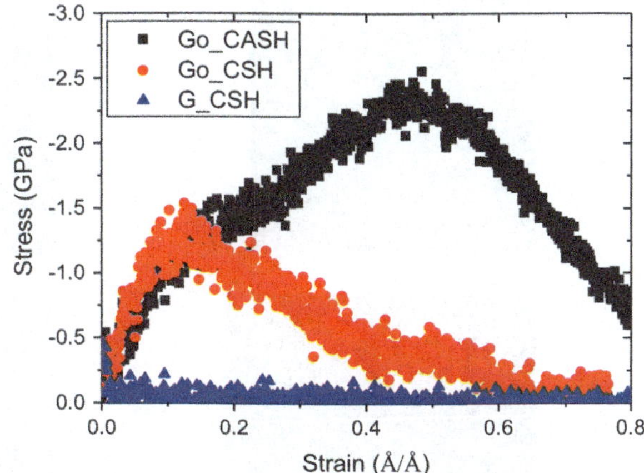

Fig. 7.11 Stress–strain relation for the G/C–S–H, GO/C–S–H, and GO/C–A–S–H tensioned along the z-direction

the G/C–S–H sample. It is worth noting that the post-failure region of the stress–strain relation (from 0.1 to 0.6 Å/Å) is elongated to some extent due to the intrusion of GO as compared with that of the tensioned C–S–H gel in previous molecular dynamics study (from 0.1 to 0.4 Å/Å) [22]. In particular, it can be observed the ladder-like stress drop at strain 0.4 Å/Å, which is a typical plasticity feature of polymers. It implies that the brittleness of the C–S–H gel is greatly improved along the interlayer direction. The uniaxial tension simulation at nanoscale level provides molecular insights on the interfacial strength between C–S–H and graphene or GO. The functionalizing greatly improved the mechanical properties of GO/cement composite because of the interfacial strength between functionalized GO nanosheets and the C–S–H gel, which further confirms the crack-bridging behavior of GO in the composite in previous experimental study.

Furthermore, the stress–strain relation in the tensioned GO/C–A–S–H sample can be divided into two stages in the stress increase process: as the strain rises to around 0.1 Å/Å, the stress–strain for the GO/C–A–S–H coincides with that of GO/C–S–H; from 0.1 to 0.5 Å/Å, the stress climbs to 2.5 GPa more slowly, with a smaller slope. Two-stage stress–strain means the reduction of the modulus after the turning point 0.1 Å/Å, but both the cohesive strength and plasticity of the C–S–H gel are significantly improved. It indicates that the Al species in the interlayer space further help in bridging the C–S–H and GO structures in order to bear the loading more effectively.

To better understand the stress–strain evolution, it is necessary to monitor the morphology changes for GO/C–S–H samples during the tension process. As shown in Fig. 7.12a, as the strain increases from 0 to 0.1 Å/Å, the structure is elongated slightly in the interlayer region due to the stretching of the Ca–OH–C bonds and the H-bonds. In the early post-failure region, bonds breakage is widely observed at the GO/C–S–H interfacial region, resulting in the birth of small cracks. At strain 0.4 Å/Å, while part of the C–OH and C–O⁻ bonds are pulled out from the C–S–H surface, some bonds are still deeply rooted between the silicate chains and surface

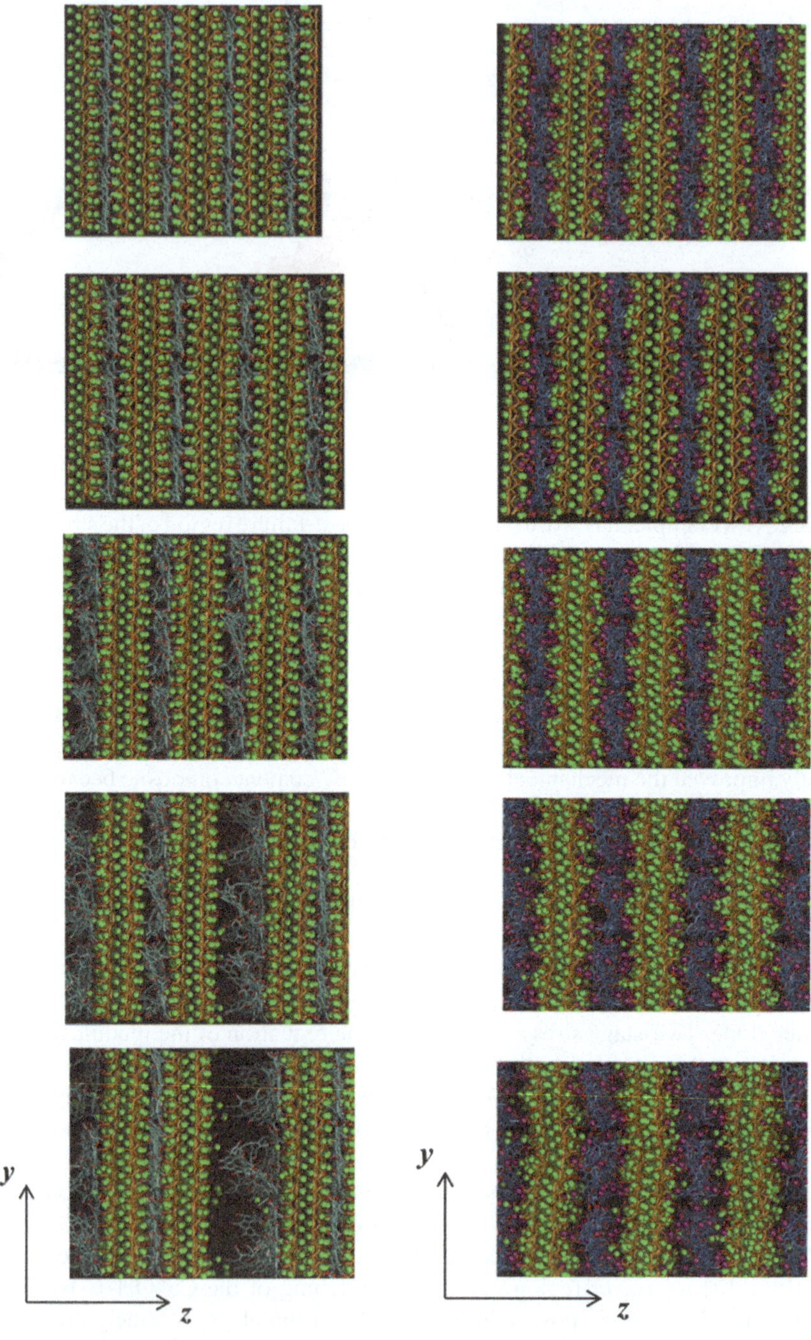

Fig. 7.12 Molecular structure evolution at z-direction tensile strain 0, 0.1, 0.2, 0.4, and 0.6 Å/Å (from top to bottom) for **a** GO/C–S–H; **b** GO/C–A–S–H composites. The meanings of the ball and sticks are illustrated in Fig. 7.1 caption

calcium atoms. The GO structure is extended along z-direction, bridging the gap between the neighboring C–S–H substrates. The GO structural extension can slow down the crack growth, prevent the sample from fracturing suddenly, and enhance the plasticity of the structure, which explains the ladder-like stress reduction in the stress–strain relation.

Compared with the morphological evolution of the GO structure, most of the interlayer calcium ions remain in the silicate surface and the calcium silicate sheet maintains an ordered layer state during the tension process. Additionally, the structural evolution of the GO/C–A–S–H is shown in Fig. 7.12b. There are two obvious structural discrepancies for C–S–H and C–A–S–H during tension process. First, at the later stage of the stress rising from 0.1 to 0.4 Å/Å, the straight arrangement of the silicate chains in the layered structure is disturbed to some extent, and the deformation of the calcium silicate sheet and GO happens simultaneously. It means that damage occurs not just at the interface but is also transferred to the C–S–H skeleton. The deformation transference also indicates rearrangement of the GO structure and is further discussed in the next section. Another difference is that the Al species is distributed through the interlayer region, not just at the interfacial region during the tensile process. Since Al atoms can continue healing the damaged structures, the cracks in the GO/C–A–S–H sample develop quite slowly.

The Al species play an important role in bridging the C–S–H gel and the GO, and help construct a network structure in the GO/C–A–S–H sample. In the GO/C–A–S–H, as shown in Fig. 7.13a, b, Q_0 species and Q_1 species correspond to C–OH/C–O groups without bonding and binding to the Al atom, respectively. The Al species, linking GO and C–S–H, has a more polymerized degree and is in the form of Q_2 and Q_3. Meanwhile, the bridging silicate tetrahedron connected with aluminate tetrahedron also forms the Q_3 branch structure and the tetrahedron pair, not directly bonded with Al species and C–OH groups, remains as the Q_2 species. With increasing n, the Q species transforms from a one-dimensional chain-like structure to a three-dimensional branch with more connections to their neighbors. A previous study proposed that the Q_3 species, present in the interlayer region, can greatly enhance the stiffness of the C–S–H structure [39].

When the structure under goes tension loading, the Al–O–Si and C–O–Al bonds are stretched broken and reconstructed, resulting in the Q species evolution. As shown in Fig. 7.14, when the strain increases from 0 to 0.4 Å/Å (corresponding to the stress rising stage in the stress–strain relation), the Q_3 and Q_1 species gradually decrease and Q_2 continuously decreases. In the post-failure stage, the Q species vary slightly, with some fluctuations. The Q_3 reduction and Q_2 increase are mainly attributed to the breakage between the Al species and the bridging silicate tetrahedron. The Q_1 reduction is partly due to the fracture between the Al species and the C–OH or C–O groups. It confirms that the Al species plays a bridging role in withstanding the loading. Interestingly, the Q_0 species remains unchanged during the first 0.2 Å/Å and continues decreasing from 0.2 to 0.4 Å/Å. The reduction of Q_0 species means that the C–OH or C–O can reconnect with the neighboring species, producing a C–O–Al connection. It should be noted that the turning point at 0.2 Å/Å agrees well with that observed in the stress–strain relation. When the structure is locally stretched broken,

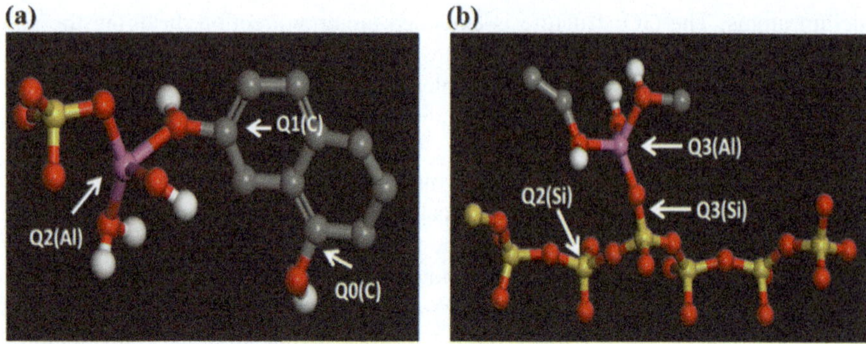

Fig. 7.13 Molecular structure of the **a** Q_0, Q_1, Q_2; and **b** Q_3 species

Fig. 7.14 Evolution of Q species percentage with the strain for GO/C–A–S–H sample

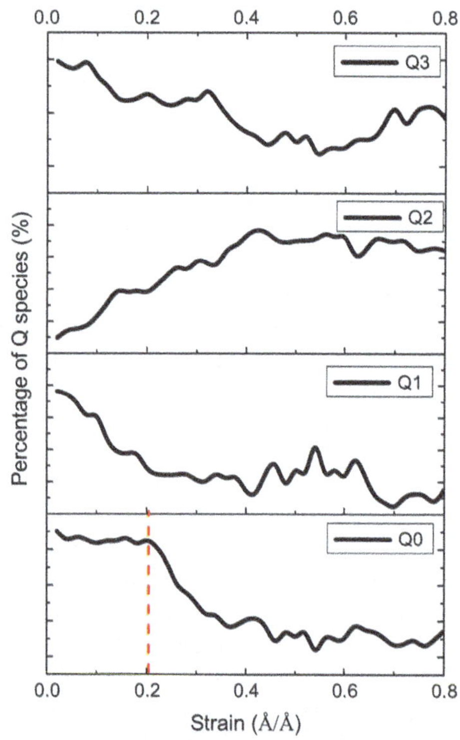

Al atoms can diffuse into the broken region and heal the structure to some extent. Therefore, a healed network can withstand more tensile strain and can enhance the plasticity of the GO/C–A–S–H sample.

7.6 Chapter Summary

The intrinsic interaction between the graphene/GO and cement hydrate in high-performance cement-based composites has been investigated by reactive molecular dynamics simulation at the atomic levels. The following conclusions were drawn from the investigations.

1. The hydrophilic nature of the interface region between the GO sheet and C–S–H is mainly attributed to the fact that the silicate chains and functional hydroxyl groups in GO provide non-bridging oxygen sites, accepting hydrogen bonds of the interlayer water molecules. Besides the H-bond connections, the Ca^{2+} and Al^{3+} ions near the surface of C–S–H play a mediating role in bridging the oxygen atoms in the silicate chains and hydroxyl groups in the GO. In particular, the Al^{3+} ions both increase the silicate chain length and heal a defective GO network, constructing the silicate–aluminate–carbon skeleton in the interface region.

2. The functional hydroxyl groups can stabilize the atoms in both C–S–H and GO structures. The aluminate–silicate chains, calcium ions and C–OH groups construct the cages, strongly preventing the free diffusion of the interface water molecules. Additionally, the lifetime of the H-bonds accepted by oxygen-containing GO is longer than that of H-bonds formed by the structural water molecules in C–S–H. On the contrary, the less friction graphene interior surface, with the hydrophobic benzene-ring structures, accelerates the transport process of the water molecules, reducing the binding stability between the C–S–H and graphene system.

3. The mechanical strength of the C–S–H and graphene/GO composites is greatly influenced by the functional hydroxyl groups and the mediating counterions. While the high cohesive force and enhance the plasticity of GO-reinforced cement composites are mainly contributed by the H-bonds and covalent–ionic bonds (O–Ca–O or O–Al–O), the weakest mechanical behavior is attributed to poor bonding and the instability of the atoms in the interface region. This matches well with the experimental findings and clearly explains the discrepancy in the reinforcement mechanism between the GO and graphene. More importantly, during the uniaxial tensile simulation, the Al atoms can reconnect with the neighboring C–OH or C–O groups, slowing down the crack propagation and enhancing the plasticity of the structures.

References

1. Bunch, J. S., Van Der Zande, A. M., Verbridge, S. S., Frank, I. W., Tanenbaum, D. M., Parpia, J. M., et al. (2007). Electromechanical resonators from graphene sheets. *Science, 315*(5811), 490–493.

2. Rafiee, M. A., Rafiee, J., Wang, Z., Song, H., Yu, Z. Z., & Koratkar, N. (2009). Enhanced mechanical properties of nanocomposites at low graphene content. *ACS Nano, 3*(12), 3884–3890.

3. Li, C., Adamcik, J., & Mezzenga, R. (2012). Biodegradable nanocomposites of amyloid fibrils and graphene with shape-memory and enzyme-sensing properties. *Nature Nanotechnology, 7*(7), 421–427.

4. Wicklein, B., Kocjan, A., Salazar-Alvarez, G., Carosio, F., Camino, G., Antonietti, M., et al. (2015). Thermally insulating and fire-retardant lightweight anisotropic foams based on nanocellulose and graphene oxide. *Nature Nanotechnology, 10*(3), 277–283.

5. Lee, C., Wei, X. D., Kysar, J. W., & Hone, J. (2008). Measurement of the elastic properties and intrinsic strength of monolayer graphene. *Science, 321*(5887), 385–388.

6. Li, Z. J. (2011). *Advanced concrete technology*. Wiley.

7. Georgakilas, V., Tiwari, J. N., Kemp, K. C., Perman, J. A., Bourlinos, A. B., Kim, K. S., et al. (2016). Noncovalent functionalization of graphene and graphene oxide for energy materials, biosensing, catalytic, and biomedical applications. *Chemical Reviews, 116*(9), 5464–5519.

8. Pan, Y., Wu, T., Bao, H., & Li, L. (2011). Green fabrication of chitosan films reinforced with parallel aligned graphene oxide. *Carbohydrate Polymers, 84*(4), 1908–1915.

9. Lv, S., Ma, Y., Qiu, C., Sun, T., Liu, J., & Zhou, Q. (2013). Effect of graphene oxide nanosheets of microstructure and mechanical properties of cement composites. *Construction and Building Materials, 49,* 121–127.

10. Pan, Z., He, L., Qiu, L., Korayem, A. H., Li, G., Zhu, J. W., et al. (2015). Mechanical properties and microstructure of a graphene oxide–cement composite. *Cement & Concrete Composites, 58,* 140–147.

11. Lu, Z., Hou, D., Meng, L., Sun, G., Lu, C., & Li, Z. (2015). Mechanism of cement paste reinforced by graphene oxide/carbon nanotubes composites with enhanced mechanical properties. *RSC Advances, 5*(122), 100598–100605.

12. Abrishami, M. E., & Zahabi, V. (2016). Reinforcing graphene oxide/cement composite with NH_2 functionalizing group. *Bulletin of Materials Science, 39*(4), 1–6.

13. Alkhateb, H., Al-Ostaz, A., Cheng, H. D., & Li, X. (2013). Materials genome for graphene–cement nanocomposites. *Journal of Nanomechanics & Micromechanics, 3*(3), 67–77.

14. Sanchez, F., & Zhang, L. (2008). Molecular dynamics modeling of the interface between surface functionalized graphitic structures and calcium–silicate–hydrate: Interaction energies, structure, and dynamics. *Journal of Colloid and Interface Science, 323*(2), 349–358.

15. Peyvandi, A., Soroushian, P., Abdol, N., & Balachandra, A. M. (2013). Surface-modified graphite nanomaterials for improved reinforcement efficiency in cementitious paste. *Carbon, 63*(2), 175–186.

16. Chenoweth, K., van Duin, A. C. T., & Goddard, W. A. (2008). ReaxFF reactive force field for molecular dynamics simulations of hydrocarbon oxidation. *Journal of Physical Chemistry A, 112*(5), 1040–1053.

17. van Duin, A. C. T., Strachan, A., Stewman, S., Zhang, Q., Xu, X., & Goddard, W. A. (2003). ReaxFFsio reactive force field for silicon and silicon oxide systems. *The Journal of Physical Chemistry A, 107*(19), 3803–3811.

18. Manzano, H., Moeini, S., Marinelli, F., van Duin, A. C. T., Ulm, F. J., & Pellenq, R. J. M. (2011). Confined water dissociation in microporous defective silicates: Mechanism, dipole distribution, and impact on substrate properties. *Journal of the American Chemistry Society, 134*(4), 2208–2215.

19. Hamid, S. (1981). The crystal structure of the 11 A natural tobermorite $Ca_{2.25}Si_3O_{7.5}(OH)_{1.5}H_2O$. *Zeitschrift fur Kristallographie, 154,* 189–198.

20. Montes-Morán, M. A., van Hattum, F. W. J., Nunes, J. P., Martínez-Alonso, A., Tascón, J. M. D., & Bernardo, C. A. (2005). A study of the effect of plasma treatment on the interfacial properties of carbon fibre–thermoplastic composites. *Carbon, 43*(8), 1795–1799.

21. Allington, R. D., Attwood, D., Hamerton, I., Hay, J. N., & Howlin, B. J. (1998). A model of the surface of oxidatively treated carbon fibre based on calculations of adsorption interactions with small molecules. *Composites Part A: Applied Science and Manufacturing, 29*(9), 1283–1290.

22. Hou, D., Zhao, T., Ma, H., & Li, Z. (2015). Reactive molecular simulation on water confined in the nanopores of the calcium silicate hydrate gel: Structure, reactivity, and mechanical properties. *The Journal of Physical Chemistry C, 119*(3), 1346–1358.

23. Youssef, M., Pellenq, R. J. M., & Yildiz, B. (2011). Glassy nature of water in an ultraconfining disordered material: The case of calcium silicate hydrate. *Journal of American Chemistry Society, 133*(8), 2499–2510.
24. Qomi, M. J. A., Ulm, F. J., & Pellenq, R. J. M. (2012). Evidence on the dual nature of aluminum in the calcium–silicate–hydrates based on atomistic simulations. *Journal of the American Ceramic Society, 95*(3), 1128–1137.
25. Li, D., Muller, M. B., Gilje, S., Kaner, R. B., & Wallace, G. G. (2008). Processable aqueous dispersions of graphene nanosheets. *Nature Nanotechnology, 3*(2), 101–105.
26. Mead, R. N., & Mountjoy, G. (2006). A molecular dynamics study of the atomic structure of $(CaO)_x(SiO_2)_{1-x}$ glasses. *Journal of Physical Chemistry, 110*(29), 273–278.
27. Pellenq, R. J. M., Kushima, A., Shahsavari, R., Van Vliet, K. J., Buehler, M. J., & Yip, S. (2009). A realistic molecular model of cement hydrates. *PNAS, 106*(38), 16102–16107.
28. L'Hôpital, E., Lothenbach, B., Kulik, D. A., & Scrivener, K. (2016). Influence of calcium to silica ratio on aluminium uptake in calcium silicate hydrate. *Cement and Concrete Research, 85*, 111–121.
29. Hou, D. S., Li, Z. J., & Zhao, T. J. (2015). Reactive force field simulation on polymerization. *RSC Advances, 5*, 448–461.
30. Hou, D. S., & Li, Z. J. (2014). Molecular dynamics study of water and ions transported during the nanopore calcium silicate phase: Case study of jennite. *Journal of Materials in Civil Engineering, 26*(5).
31. Hou, D. S., Li, Z. J., Zhao, T. J., & Zhang, P. (2015). Water transport in the nano-pore of the calcium silicate phase: Reactivity, structure and dynamics. *Physical Chemistry Chemical Physics, 17*, 1411–1423.
32. Coudert, F.-X., Vuilleumier, R., & Boutin, A. (2006). Dipole moment, hydrogen bonding and IR spectrum of confined water. *Physical Chemistry Chemical Physics, 7*(12), 2464–2467.
33. Cai, W. W., Piner, R. D., Stadermann, F. J., Park, S., Shaibat, M. A., Ishii, Y., et al. (2008). Synthesis and solid-state NMR structural characterization of ^{13}C-labeled graphite oxide. *Science, 321*, 1815–1817.
34. Qomi, M. J. A., Bauchy, M., Ulm, F. J., & Pellenq, R. J. M. (2014). Anomalous composition-dependent dynamics of nanoconfined water in the interlayer of disordered calcium-silicates. *The Journal of Chemical Physics, 140*(5), 054515.
35. Holt, J. K., Park, H. G., Wang, Y., Stadermann, M., Artyukhin, A. B., Grigoropoulos, C. P., et al. (2006). Fast mass transport through sub-2-nanometer carbon nanotubes. *Science, 312*(5776), 1034–1037.
36. Falk, K., Sedlmeier, F., Joly, L., Netz, R. R., & Bocquet, L. (2010). Molecular origin of fast water transport in carbon nanotube membranes: Superlubricity versus curvature dependent friction. *Nano Letters, 10*(10), 4067–4073.
37. Alexiadis, A., & Kassinos, S. (2008). Molecular simulation of water in carbon nanotubes. *Chemical Reviews, 108*(12), 5014–5034.
38. Bonnaud, P. A., Ji, Q., Coasne, B., Pellenq, R. J. M., & Van Vliet, K. J. (2012). Thermodynamics of water confined porous calcium silicate hydrate. *Langmuir, 28*(31), 11422–11432.
39. Merlino, S., Bonnacorsi, E., & Armbruster, T. (2001). The real structure of tobermorite 11 A: Normal and anomalous forms, OD character and polyptic modifications. *European Journal of Mineralogy, 13*, 577–590.

Chapter 8
The Future and Development Trends of Computational Chemistry Applied in Concrete Science

In previous sections, the atomistic simulation methods have been introduced to the materials science for cement-based materials to decode the intrinsic building block of the cement hydrate at nanoscale. The molecular dynamics method exhibits the significant advantage in investigating the properties of cement-based concrete material at nanoscale and opens a novel pathway for design of construction and building materials. The MD model of C–S–H, validated by experimental studies, has been successfully applied to unravel the relationship between molecular structure and mechanical properties of the material, and help understand the structural and properties evolution as the function of chemical composition. This can provide molecular insight into guiding the high-performance material design by composition optimization. Furthermore, it has been widely utilized to study water and ions migration in the gel pores of cement hydrate, which is the critical scientific problem for concrete durability. The hydration, diffusion, and capillary transport behavior of the solution species in the vicinity of cement-hydrate surface have been systematically investigated to support the durability study. Another contribution of MD study is to reveal molecular properties of the cement hydrate incorporated with supplementary cementitious material (SCM) from the industrial mineral waste. This helps develop the environmental friendly concrete material to lower the carbon footprint of construction industry. Also, the beauty of the atomistic simulation method provides valuable theoretical basis for the application of nanotechnology in concrete material, such as the graphene oxide modified cementitious composite. Due to the unique advantage, the application of MD method in scientific field of concrete will keep increasing in the future. While the computational chemistry has been widely utilized in many fields such as bioscience, physics, and chemistry for a long time, the application in the concrete scientific field only has ten-year history. Currently, many researches based on computational chemistry are the preliminary attempts, and further investigation is needed for more accurate, systematically exploring many intrinsic mechanisms in cement-based material. The research and development of MD method suitable for the concrete study have to be advanced to satisfy the need and new requests to face new challenges. Hence, the perspectives closely related to MD applied in concrete science are briefly discussed.

© Science Press and Springer Nature Singapore Pte Ltd. 2020

D. Hou, *Molecular Simulation on Cement-Based Materials*,

https://doi.org/10.1007/978-981-13-8711-1_8

8.1 Force Field Database Development for Cement-Based Material

Force field parameterization for molecular dynamics is essential for modeling of materials at molecular scale. It determines the accuracy of various properties that are predicted by computational chemistry method, such as the molecular structure, the vibration behavior of the chemical bonds and the mechanical properties of the materials, et al. Generally, parameterization of the force field derived from ab initio study and validated by the experimental results, is the first step for the atomistic simulation of related materials. A good force field can significantly contribute to the development of the molecular simulation in their fields. For example, the COMPASS [1] force field is developed for the Condensed-phase optimized potential that enables accurate and simultaneous prediction of structural, conformational, vibrational, and thermos-physical properties for a broad range of molecules in isolation and in condensed phases. It also consolidates parameters for organic and inorganic materials previous found in different force fields. Furthermore, development of the force field needs continuous efforts in parameterization, validation and updating to improve the accuracy, transferability, compatibility, and stability of potentials. In addition to mainly utilized force fields (ClayFF, CSHFF and ReaxFF) in previous chapters, many force fields have been developed for cement system, including BMH (Born—Mayer–Huggins) [2], InterfaceFF (IFF) [3], CementFF [4], and UFF [5]. Generally speaking, all the force fields parameterization are most reliable for the behaviors that they have been validated with experimental results and from ab initio calculations. It is technically difficult for one single approach to characterize well all the features of the cement hydrate at molecular level. The force fields have their advantages and limitation. ClayFF accurately describes the bulk structures of a broad range of simple hydroxide and oxyhydroxides phases. Parametrization of ClayFF using a set of simple structurally well-characterized hydrated phases allows good transferability of the force field parameters. ClayFF is an easily implemented and computationally low-cost force field which probably needs some reparametrization to make it suitable for further more ambitious applications. It has been found that modeling the mineral analogues of cement hydrate, such as tobermorite by ClayFF underestimates the mechanical properties of the layered crystal. It means that transferability of force fields to analogous hydrated oxides without rigorous investigations may result in misleading property predictions. To remedy the deficiency of single point charge ClayFF force field, CSHFF has been fitted to many ab initio-based lattice parameters and elastic constants of a series of cement-hydrate minerals. Hence, mechanical properties predicted by CSHFF match well with the experimental and first principle calculation results. The principal limitation of non-reactive classical force fields, ClayFF and CSHFF, is that they cannot simulate the chemical reactions including the chemical bond formation and dissociation. The ReaxFF force field overcomes the limitation and reproduces accurately the energy evolution induced by the bond formation and dissociation. It makes modeling the chemical reaction in the cement system available and also helps understand the hydration process of the cement mineral, the fracture mechanism of cement hydrate subject to large deformation, and the reactivity of water ultra-confined in nanometer gel pores. Specialized energy expressions of intermolecular potentials in ReaxFF are typically limited in compatibility with other harmonic energy expressions such as COMPASS and CVFF. In this scenario, it is difficult to use interatomic potential parameters in ReaxFF to study the interactions between organic material and cementitious minerals. Recently, the *cemff* database [6], developed by many researchers in molecular modeling of cement system, has been established and can be accessible via web-link (http://cemff.epfl.ch). The *cemff* database can help researchers to evaluate and choose the suitable potential for specific system. The database can also be updated by incorporating newly developing force field and modifying the inaccurate part in previous force field.

8.2 Mesoscale Modeling of the Cement Hydrate by Coarse Grain Molecular Dynamics

Although MD method is a powerful tool in dealing with the nanoscale simulation of cementitious materials in both chemical and mechanical properties, the nanoscale simulation results cannot be directly put into practical or large-scale experimental use due to the limitation in both time and length scales. In order to upscale the study of C–S–H to a higher level without neglecting the material's essential properties in the nanoscale, upscaled simulation in mesoscale is becoming necessary.

Coarse grain molecular dynamics (CGMD) provides a pathway to bridge the gap between molecular simulation and microlevel modeling. The coarse grain methods have been widely utilized in modeling the macromolecules. As shown in Fig. 8.1, the long polymer chains can be simplified to ball and spring by containing more atoms in one CG sites. Multiscale coarse-grained (MS-CG) method was proposed by Izvekov and Voth in simulating the dimyristoylphosphatidylcholine (DMPC) lipid bilayer [7]. The parameter of this MS-CG method is derived from the atomistic (or molecular) simulation through numerical fitting and matching method. Hence, this "multiscale" represents the upscale process from MD to CG. By utilizing the variational principle in obtaining the optimized force field of this CG model directly from the modeling of MD method inner the same system [8, 9], the derived potential of mean force (PMF) is capable of describing the interactions of CG sites accurately in accordance with MD method while reducing the calculation expansion at the same time. However, since the CG force field is obtained under the same system with MD method, it is not simple for wider application under other circumstances. Besides, the parameterization process of variational process can be computation-demanding at some time. These two shortcomings constrain a wider application of MS-CG simulation method.

In addition to the MS-CG method mentioned above in solving the CG force field problem, another solution for the problem of CG method is to include more molecules inner one CG-molecule particle, three for example [10], thus reducing

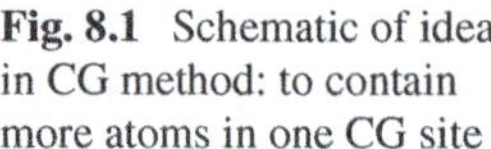

Fig. 8.1 Schematic of idea in CG method: to contain more atoms in one CG site

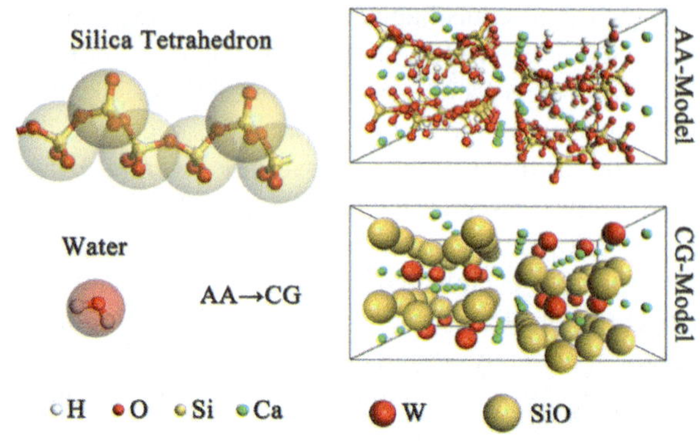

the polarity of torsion. Similarly, another CG method, called effective force coarse-graining (EF-CG) method, is proposed to determine the pairwise interaction between two CG sites [11]. This EF-CG method treats all the CG sites, no matter bonded or un-bonded, separately as the MD method does for the single atom. Then, different types of EF-CG force field are determined for different pairs of atoms, where one CG site is regarded in the center-of-mass (COM) of the atoms included in it. Since the EF-CG method is an upscaling method from MD method, it can only be applied to systems with the utilization of molecular simulation. Since there are more general assumptions, EF-CG method is less accurate as the MS-CG method, but it is more transferable than the later.

For the complex cement-hydrate system, the C–S–H model at mesoscale has been widely accepted as nanoparticles aggregation from experimental investigation and theoretical modeling. Interaction forces between these nanoparticles are at the origin of C–S–H chemical, physical, and mechanical properties at the mesoscales. These particle interactions and the resulting properties may be affected significantly by nanoparticle density and environmental conditions such as the temperature, relative humidity, or concentration of chemical species in the bulk solution. Bonnaud investigated the pair potential between the clusters of box-shaped C–S–H by grand canonical Monte Carlo and mean force integration method under various configurations [12].

Based on the simplification idea CG method, disk-like C–S–H simplification is put forward in simulating the mesoscopic properties with different packing types [13, 14]. But the disk-shaped C–S–H may cause some disorder in the packing process due to the significant difference in three dimensions, while the generally accepted idea that C–S–H can be regarded as spherical colloidal particles in the mesoscale would avoid this kind of instability [15]. As shown in Fig. 8.2, combined with dynamic simulation, the packing idea [16, 17] of colloidal spheres in modeling the precipitation process is adopted in the mesoscale simulation Tnend the mechanical performance like stress–strain relationship and structural characteristics such as pore size distribution have been well explored. On the other hand, C–S–H is an anisotropic material in the aspect of mechanics under nanoscale. But in larger scales, like microscale, cement paste represents a more isotropic mechanical behavior [18] due to a relatively

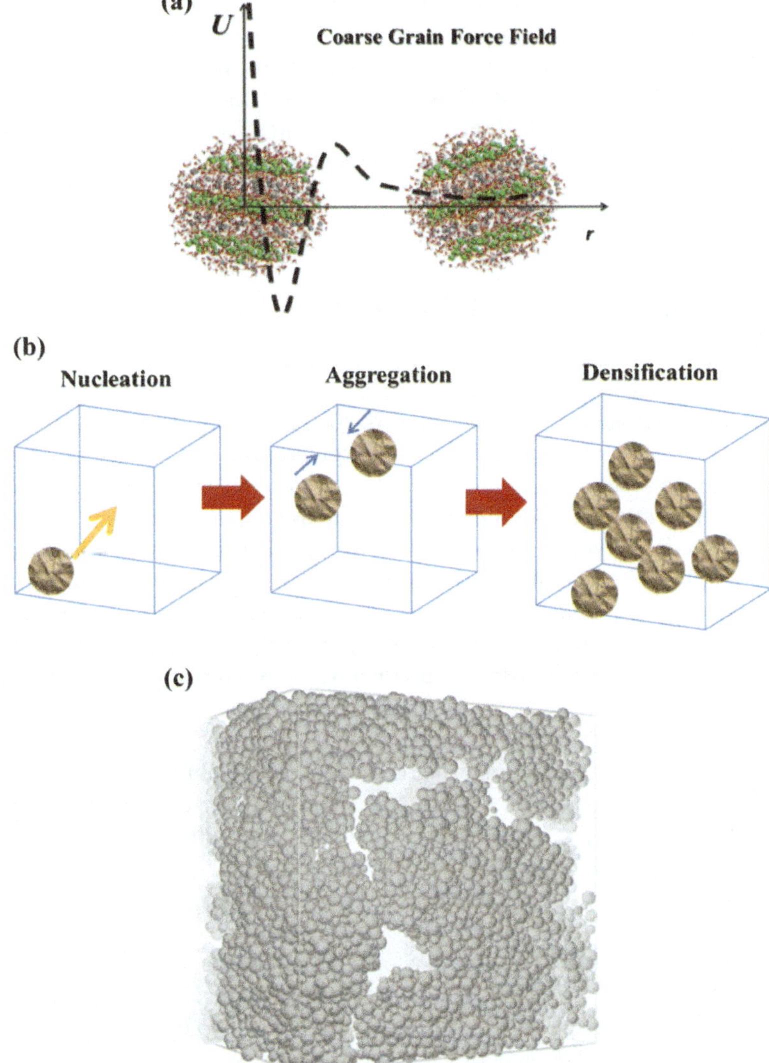

Fig. 8.2 **a** Schematic illustration of coarse-grained force field for cement hydrate; **b** GCMC method study on the nanoparticles growth during hydration process; **c** the mesoscale model of C–S–H gel

uniform distribution in components. It would be computation consuming to directly upscale the anisotropic atomistic modeling to isotropic simulation in mechanics. The simplification idea of colloidal composition [19] would act as a feasible way in linking both the accuracy and simplification purpose. The major task of the future work is to establish a theoretical model for C–S–H in a larger scale over nanoscale, which could behave like a bridge to link the nanoscale (1–10 nm) and microscale (~100 μm).

8.3 Molecular Modeling of Low Carbon Geopolymer Binders

To reduce the environmental impact of Portland cement, efforts have been made to search for other types of binders with low energy demand and less emission of carbon dioxide. One system along this line is alkaline-activated cementitious materials, in which geopolymer and alkaline-activated slag are two good examples. It is well known that the geopolymer is composed of NASH gel. There is a striking similarity between NASH gel and sodium aluminosilicate glass. According to the results of spectroscopy [20–24], both are composed of rings and chains structure of Si and Al tetrahedron interconnected together to form a threedimensional network structure. It is quite different from the hydration product (C–A–S–H) from cement and slag in Chap. 6. In the Al-rich cement hydrate, the long silicate-aluminate chain occupies predominated percentage instead of branch structures. In N–A–S–H, the presence of AlO_4 tetrahedron creates a negative charge imbalance for aluminosilicate skeleton, where monovalent cations such as K^+ or Na^+ are immobilized for compensating the charge. Additionally, there are also other techniques contributing to develop a molecular understanding of the chemical properties of geopolymer, including calorimetry [25], X-ray and neutron scattering [26, 27], and neutron pair distribution function (PDF) analysis [28].

Although a considerable amount of molecular modeling work has been undertaken to understand the atomic level structure and mechanical property interplay of CSH gel in previous chapters, a very limited number of studies have been carried out to examine the intrinsic geopolymer binder phase. White et al. [29] used density functional theory (DFT)-based coarse-grained Monte Carlo (CGMC) simulation technique to understand the molecular mechanisms responsible for the structural changes that occur during geopolymerization using metakaolin as source materials. Due to the similarity with silicate-aluminate glass, the molecular model of amorphous geopolymer binder was constructed by heating and cooling cycles utilized in the glass production process in previous studies. The molecular structure of NASH gel after quenching simulation and water adsorption was shown in Fig. 8.3. Water molecules penetrate into the cavity region among aluminosilicate skeleton and are irregularly distributed inside the NASH gel. An atomic-scale analysis showed a direct correlation between water content and diffusion of alkali ions, resulting in the weak-

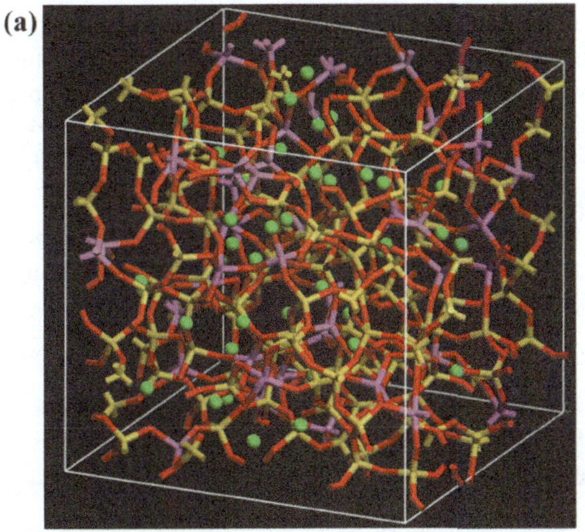

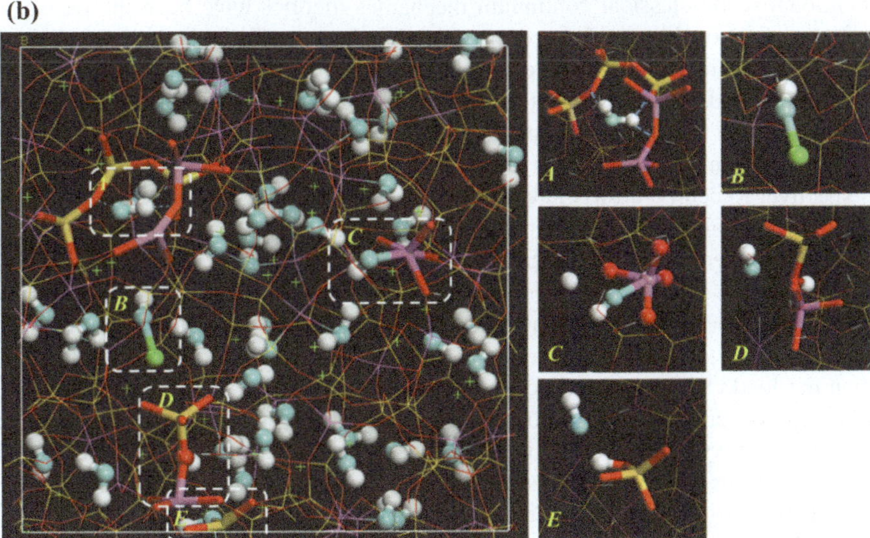

Fig. 8.3 **a** Molecular model of geopolymer; **b** water adsorption and dissociation in the silicate-aluminate skeleton

ening of the AlO_4 tetrahedral structure due to the migration of charge balancing alkali ions away from the tetrahedra, ultimately leading to failure. Molecular dynamics can thus serve as a useful design tool for optimizing the composition of geopolymers with improved mechanical properties.

Furthermore, many theories in the glass science can be borrowed to investigate the molecular structural and mechanical properties. For example, rigidity theory is a powerful tool to predict the properties of glasses with respect to composition [30].

By reducing such molecular networks to simple mechanical trusses, topological constraints theory filters out all the unnecessary details that ultimately do not affect macroscopic properties. The usual constraints enumeration is restricted to networks that are amorphous, homogeneous, and fully connected. How rigidity theory can be used to describe the nanoscale structure of this geopolymer material, by relying on molecular dynamics simulations. The distinction between intact and broken constraints should be clearly defined at the atomic scale, thus allowing a precise enumeration of the topological constraints. The rigidity theory can help improve the hardness of geopolymer materials.

8.4   Solutions of Concrete Structural Engineering from Molecular Dynamics

Traditionally, the classical continuum mechanics theories have been the basis for most computational methods used in various engineering fields including civil and mechanical engineering; examples are finite elements, finite difference, finite volume, and boundary element methods. The capability of the continuum approach is limited when structural solution at a small length scale is of concern, or if predictions about material behavior should be made from a fundamental bottom-up perspective. Recently, efforts have been made toward applying the MD method to structural mechanics and engineering. One typical example of the application of MD in Civil Engineering is to study the interfacial bond between FRP (Fiber Reinforcement Polymer) and concrete. In practice, FRP-strengthened system for a RC beam is typically designed such that failure occurs in the form of either steel reinforcement yielding followed by concrete crushing, or FRP rupture, providing early warning when the load capacity is exceeded. However, when the interface becomes weak, failure can occur in the form of delamination in a tri-layer material system formed by the FRP, epoxy adhesive (organic), and the concrete material (inorganic) at various critical locations along the beam soffit, leading to a significantly lower load capacity of the FRP-retrofitted system. Figure 8.4 shows possible debonding configurations in a FRP-retrofitted reinforced concrete beam. Existing knowledge on mesoscale fracture mechanics may not fully explain the weakening of the interface between concrete and epoxy, when the interface is under moisture; there is a need to study the moisture affected debonding of the interface using a more fundamental approach that incorporates chemistry in the description of materials. The interfacial model of epoxy polymer and silica has been established in moisture and dry condition in recent study [31]. The results of the atomistic modeling showed that the adhesive strength (in terms of energy) between epoxy and silica is weakened in the presence of water through its interaction with epoxy. This is correlated with the existing mesoscale experimental data. Even though the silica cannot completely represent the concrete surface structure, the example gives first successful attempt on solving the structural engineering problem. This example demonstrates that MD simulation can be effec-

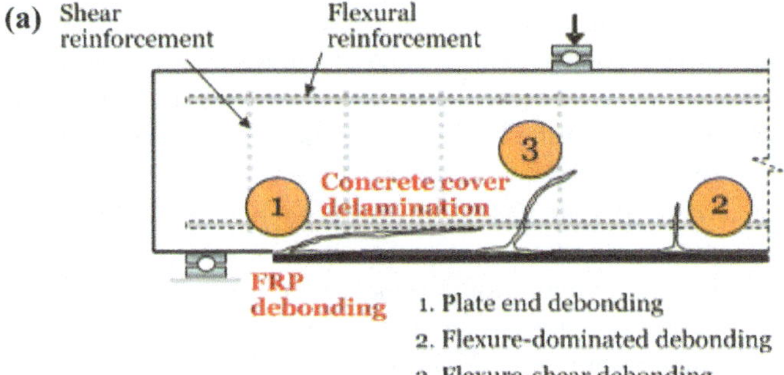

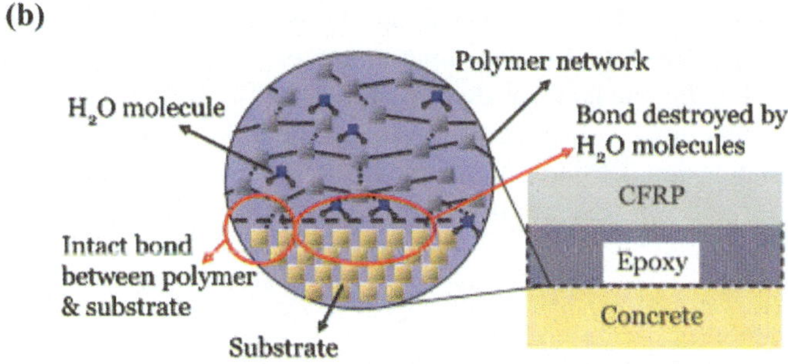

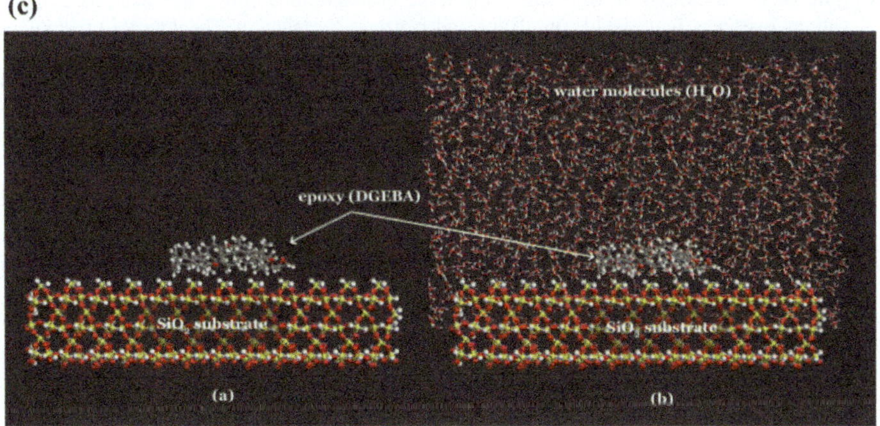

Fig. 8.4 **a** Typical failure modes of FRP-retrofitted reinforced concrete beam; **b** weakening of bond between epoxy and concrete; **c** simulation model for **a** dry case and **b** wet case [31]

tively used in studying the durability of the system through an understanding of how materials interact with the environment at the molecular level. It should be noted that the interfacial bond problem is a critical scientific issue for structural engineering. In civil engineering, the failure of the RC structural element frequently happens at the interface between concrete and reinforcement steel. The intrinsic interaction between passive film and cement hydrate has not been fully understood. Also, as discussed in Chap. 5, detrimental ions and water migrate into the concrete by the pores, resulting in the degradation of the hydration product and corrosion of the reinforcement steel. In order to inhibit the ingress of the detrimental ions, the hydrophobic coating material is always incorporated on the surface of concrete structure. The cohesive strength between the coating film and cement hydrate determines the effectiveness of the water resistant ability. Generally speaking, the weakness of RC structure exists in the brittle and porous nature of the concrete material and the degradation of the interfacial bond. On the one hand, as discussed in Chap. 4, the MD can be employed to study the compositional influence on the mechanical properties of the C–S–H to guide the material design. Besides, it also can study the interfacial bond mechanism.

References

1. Sun, H. (1998). COMPASS: An ab initio force-field optimized for condensed-phase applications overview with details on alkane and benzene compounds. *The Journal of Physical Chemistry B, 102*(38), 7338–7364.
2. Faucon, P., Delaye, J. M., Virlet, J., Jacquinot, J. F., & Adenot, F. (1997). Study of the structural properties of the C–S–H(I) by molecular dynamics simulation. *Cement and Concrete Research, 27*(10), 1581–1590.
3. Heinz, H., Lin, T.-J., Kishore Mishra, R., & Emami, F. S. (2013). Thermodynamically consistent force fields for the assembly of inorganic, organic, and biological nanostructures: The INTERFACE force field. *Langmuir, 29*(6), 1754–1765.
4. Galmarini, S., Aimable, A., Ruffray, N., & Bowen, P. (2011). Changes in portlandite morphology with solvent composition: Atomistic simulations and experiment. *Cement and Concrete Research, 41*(12), 1330–1338.
5. Rappe, A. K., Casewit, C. J., Colwell, K. S., Goddard, W. A., & Skiff, W. M. (1992). UFF, a full periodic table force field for molecular mechanics and molecular dynamics simulations. *Journal of the American Chemical Society, 114*(25), 10024–10035.
6. Mishra, R. K., Mohamed, A. K., Geissbühler, D., Manzano, H., Jamil, T., Shahsavari, R., et al. (2017). cemff: A force field database for cementitious materials including validations, applications and opportunities. *Cement and Concrete Research, 102*, 68–89.
7. Izvekov, S., & Voth, G. A. (2005). A multiscale coarse-graining method for biomolecular systems. *Journal of Physical Chemistry B, 109*(7), 2469–2473.
8. Noid, W. G., Chu, J. W., Ayton, G. S., Krishna, V., Izvekov, S., Voth, G. A., et al. (2008). The multiscale coarse-graining method. I. A rigorous bridge between atomistic and coarse-grained models. *The Journal of Chemical Physics, 128*(24), 2469.
9. Noid, W. G., Liu, P., Wang, Y., Chu, J. W., Ayton, G. S., Izvekov, S., et al. (2008). The multiscale coarse-graining method. II. Numerical implementation for coarse-grained molecular models. *Journal of Chemical Physics, 128*(24), 2469–2307.
10. Shinoda, W., DeVane, R., & Klein, M. L. (2008). Coarse-grained molecular modeling of nonionic surfactant self-assembly. *Soft Matter, 4*(12), 2454–2462.

11. Wang, Y., Noid, W. G., Liu, P., & Voth, G. A. (2009). Effective force coarse-graining. *Physical Chemistry Chemical Physics, 11*(12), 2002–2015.
12. Bonnaud, P. A., Labbez, C., Miura, R., Suzuki, A., Miyamoto, N., Hatakeyama, N., et al. (2016). Interaction grand potential between calcium–silicate–hydrate nanoparticles at the molecular level. *Nanoscale, 8*(7), 4160–4172.
13. Yu, Z., & Lau, D. (2015). Nano- and mesoscale modeling of cement matrix. *Nanoscale Research Letters, 10*(1), 1–6.
14. Yu, Z., Zhou, A., & Lau, D. (2016). Mesoscopic packing of disk-like building blocks in calcium silicate hydrate. *Scientific Reports, 6*, 36967.
15. Jennings, H. M. (2008). Refinements to colloid model of C–S–H in cement: CM-II. *Cement and Concrete Research, 38*(3), 275–289.
16. Ioannidou, K., Krakowiak, K. J., Bauchy, M., Hoover, C. G., Masoero, E., Yip, S., et al. (2016). Mesoscale texture of cement hydrates. *Proceedings of the National Academy of Sciences of the United States of America, 113*(8), 2029.
17. Ioannidou, K., Del Gado, E., Ulm, F.-J., & Pellenq, R. J.-M. (2017). Inhomogeneity in cement hydrates: Linking local packing to local pressure. *Journal of Nanomechanics and Micromechanics, 7*(2), 04017003.
18. Qian, Z., Ye, G., Schlangen, E., & Van Breugel, K. (2011). 3D lattice fracture model: Application to cement paste at microscale. *Key Engineering Materials, 452–453*, 65–68.
19. Ioannidou, K., Pellenq, R. J., & Del, G. E. (2014). Controlling local packing and growth in calcium–silicate–hydrate gels. *Soft Matter, 10*(8), 1121–1133.
20. Duxson, P., Lukey, G. C., Separovic, F., & Deventer, J. S. J. V. (2005). Effect of alkali cations on aluminum incorporation in geopolymeric gels. *Industrial and Engineering Chemistry Research, 44*(4), 832–839.
21. Barbosa, V. F. F., Mackenzie, K. J. D., & Thaumaturgo, C. (2000). Synthesis and characterisation of materials based on inorganic polymers of alumina and silica: Sodium polysialate polymers. *International Journal of Inorganic Materials, 2*(4), 309–317.
22. Duxson, P., Provis, J. L., Lukey, G. C., Separovic, F., & Deventer, J. S. J. V. (2005). Si-29 NMR study of structural ordering in aluminosilicate geopolymer gels. *Langmuir, 21*(7), 3028–3036.
23. Singh, P. S., Trigg, M., Burgar, I., & Bastow, T. (2005). Geopolymer formation processes at room temperature studied by 29Si and 27Al MAS-NMR. *Materials Science and Engineering: A, 396*(1), 392–402.
24. Duxson, P., Lukey, G. C., & Deventer, J. S. J. V. (2007). Physical evolution of Na-geopolymer derived from metakaolin up to 1000 °C. *Journal Materials Science, 42*(9), 3044–3054.
25. Zhang, Z., Wang, H., Provis, J. L., Bullen, F., Reid, A., & Zhu, Y. (2012). Quantitative kinetic and structural analysis of geopolymers. Part 1. The activation of metakaolin with sodium hydroxide. *Thermochim. Acta, 539*(13), 23–33.
26. Bell, J. L., Sarin, P., Driemeyer, P. E., Haggerty, R. P., Chupas, P. J., & Kriven, W. M. (2008). X-ray pair distribution function analysis of a metakaolin-based, $KAlSi_2O_6 \cdot 5.5H_2O$ inorganic polymer (geopolymer). *Journal of Materials Chemistry, 18*(48), 5974–5981.
27. White, C. E., Provis, J. L., Proffen, T., & Deventer, J. S. J. V. (2010). The effects of temperature on the local structure of metakaolin-based geopolymer binder: A neutron pair distribution function investigation. *Journal of the American Ceramic Society, 93*(10), 3486–3492.
28. White, C. E., Provis, J. L., Llobet, A., Proffen, T., & Deventer, J. S. J. V. (2011). Evolution of local structure in geopolymer gels: An in situ neutron pair distribution function analysis. *Journal of the American Ceramic Society, 94*(10), 3532–3539.
29. White, C. E., Provis, J. L., Proffen, T., & van Deventer, J. S. J. (2012). Molecular mechanisms responsible for the structural changes occurring during geopolymerization: Multiscale simulation. *AIChE Journal, 58*(7), 2241–2253.
30. Bauchy, M., Abdolhosseini Qomi, M. J., Bichara, C., Ulm, F.-J., & Pellenq, R. J. M. (2014). Nanoscale structure of cement: Viewpoint of rigidity theory. *The Journal of Physical Chemistry C, 118*(23), 12485–12493.
31. Büyüköztürk, O., Buehler, M. J., Lau, D., & Tuakta, C. (2011). Structural solution using molecular dynamics: Fundamentals and a case study of epoxy-silica interface. *International Journal of Solids and Structures, 48*(14), 2131–2140.

The manufacturer's authorised representative in the EU is Springer
Nature Customer Service Centre GmbH, Europaplatz 3, 69115 Heidelberg,
Germany. If you have any concerns regarding our products, please
contact ProductSafety@springernature.com

Printed and bound by CPI Group (UK) Ltd, Croydon, CR0 4YY

28/11/2025

02007714-0006